AF581864

Novel Heterogeneous Catalysts for Advanced Oxidation Processes (AOPs)

Novel Heterogeneous Catalysts for Advanced Oxidation Processes (AOPs)

Editors

Olívia Salomé G.P. Soares
Carla Orge
Raquel Pinto Rocha

MDPI • Basel • Beijing • Wuhan • Barcelona • Belgrade • Manchester • Tokyo • Cluj • Tianjin

Editors

Olívia Salomé G.P. Soares
University of Porto
Portugal

Carla Orge
University of Porto
Portugal

Raquel Pinto Rocha
University of Porto
Portugal

Editorial Office
MDPI
St. Alban-Anlage 66
4052 Basel, Switzerland

This is a reprint of articles from the Special Issue published online in the open access journal *Catalysts* (ISSN 2073-4344) (available at: https://www.mdpi.com/journal/catalysts/special_issues/Advanced_Oxidation_Processes).

For citation purposes, cite each article independently as indicated on the article page online and as indicated below:

LastName, A.A.; LastName, B.B.; LastName, C.C. Article Title. *Journal Name* **Year**, *Volume Number*, Page Range.

ISBN 978-3-0365-4283-6 (Hbk)
ISBN 978-3-0365-4284-3 (PDF)

Contents

About the Editors

Olívia Salomé G.P. Soares

Olívia Salomé G.P. Soares concluded her PhD in December 2010, and was awarded for the best PhD thesis. Currently, she is an Assistant Researcher and her background is in the field of Chemical and Environmental Engineering, focused on development of innovative heterogeneous catalysts for environmental technologies (water/wastewater treatments, abatement of air pollutants), CO2 methanation, electrochemical energy conversion, and smart textiles. She has co-authored 3 book chapters, 121 papers (25 as first author, 20 as corresponding author, 73 in the last 5 years) in international peer-reviewed journals, H- index=29 and ¿2450 citations, and has had two national patents and one international patent granted. She has several national and international collaborations in several fields. She is the principal investigator (PI) of one national (SmartOxidation) and one international (NanoCatRed) research projects.

Carla Orge

Carla Orge operates in the research areas of Engineering Sciences and Technologies with an emphasis on Chemical Engineering, in the scope of environmental catalysis. She concluded her Ph.D. degree in 2012 with the thesis *Tertiary Treatment of Effluents by Catalytic Ozonation*. After that, she started a post-doc grant in the area of *Photocatalytic ozonation for micropollutants abatement in water and wastewater*. In 2018, she earned a contract as Junior Researcher. She is PI and co-PI of two ongoing projects, InTreat and SmartOxidation. She has accumulated experience as evidenced by the 28 published scientific papers in ISI journals (with almost 700 citations; h-index=15).

Raquel Pinto Rocha

Raquel Pinto Rocha graduated in Chemical Engineering and concluded her PhD in Chemical and Biological Engineering in March 2018, where she gained extensive expertise in developing carbon-based materials for liquid-phase reactions. After completing her PhD, she has worked as a post-doc Researcher to develop nanostructured based carbon materials for catalysis and energy applications. Currently, she has a contract as Junior Researcher with 26 scientific papers in ISI journals, 11 as first author, and an h-index of 15. She is also co-PI of national project (BiCat4Energy), regarding the development of carbon-based materials for renewable energy conversion.

Editorial

Novel Heterogeneous Catalysts for Advanced Oxidation Processes (AOPs)

Olívia Salomé G. P. Soares [1,2,*], Carla A. Orge [1,2] and Raquel Pinto Rocha [1,2]

1 Laboratory of Separation and Reaction Engineering—Laboratory of Catalysis and Materials, (LSRE-LCM), Faculty of Engineering, University of Porto, Rua Dr. Roberto Frias, 4200-465 Porto, Portugal; carlaorge@fe.up.pt (C.A.O.); rprocha@fe.up.pt (R.P.R.)

2 ALiCE—Associate Laboratory in Chemical Engineering, Faculty of Engineering, University of Porto, Rua Dr. Roberto Frias, 4200-465 Porto, Portugal

* Correspondence: salome.soares@fe.up.pt

With the increasing global usage of water and the continuous addition of contaminants to water sources, new challenges associated with the abatement of organic pollutants, particularly those that are refractory to conventional water and wastewater treatment technologies, have arisen. Advanced oxidation processes (AOPs) present a competitive alternative to promote the oxidation of organic contaminants by strong oxidative radicals generated from oxygen, ozone, wet peroxide, UV radiation. In addition, the use of catalysts not only improves efficiency, but may present remarkable cost advantages for practical applications of AOPs in the abatement of several pollutants.

Rocha et al. prepared N, S-co-doping of commercial carbon nanotubes (CNTs) by a solvent-free mechanothermal approach using thiourea [1]. Although the samples revealed a similar performance for phenol degradation by catalytic wet air oxidation, a higher total organic carbon removal was observed using the sample thermally treated at 900 °C, which was attributed to the presence of N6, NQ, and thiophenic groups. Martín-Somer et al. evaluated the activity of P25 TiO_2 particles, coated with SiO_2, using atomic layer deposition (ALD) during methylene blue removal by photocatalysis, oxidation of methanol and inactivation of Escherichia coli bacteria in water and compared with bare P25 [2]. A significant improvement in the removal of methylene blue is achieved, due to an increase in its adsorption. Sharif and Roberts successfully synthesized and characterized electrically conducting nanocomposites, including graphene/SnO_2 and graphene/Sb-SnO_2 [3]. The adsorption capacity of the new composites was ≥40% higher than bare graphene and were effectively regenerated in both NaCl and Na_2SO_4 electrolytes, attaining high regeneration efficiencies and no loss of the nanocomposite. A detailed study about the environmental impacts associated with the production of different magnetic nanoparticles (NPs) based on magnetite (Fe_3O_4), with a potential use as heterogeneous Fenton or photo-Fenton catalysts in wastewater treatment applications was presented by Feijoo et al. [4]. The results suggest that magnetic nanoparticles coated with stabilizing agents as poly(acrylic acid) (PAA) and polyethylenimine (PEI) were the most suitable option for their applications in heterogeneous Fenton processes, whereas ZnO-Fe_3O_4 NPs provided an interesting approach in photo-Fenton. Wang et al. verified that prepared novel biochar-supported composite containing both iron sulfide and iron oxide enhanced the catalytic degradation of ciprofloxacin through Fenton-type reactions, due to increase production of ·OH radicals [5]. Santos Silva et al. using clay-based materials in the catalytic wet peroxide oxidation during paracetamol (PCM) degradation and a high stability was observed due to lower leaching verified at the end of the reaction [6]. Orge et al. studied the presence of magnetic nanoparticles (MNP) composed of iron oxide coated with carbon in the photocatalytic ozonation and verified that the carbon phase is directly related to high catalytic activity [7]. Activated carbon (AC), carbon xerogel (XC), and carbon nanotubes (CNT), with and without N-functionalities, impregnated with iron were evaluated during adsorption, catalytic wet

Citation: Soares, O.S.G.P.; Orge, C.A.; Rocha, R.P. Novel Heterogeneous Catalysts for Advanced Oxidation Processes (AOPs). *Catalysts* **2022**, *12*, 498. https://doi.org/10.3390/catal12050498

Received: 21 April 2022
Accepted: 27 April 2022
Published: 29 April 2022

peroxidation (CWPO), and Fenton process during p-nitrophenol (PNP) degradation by Soares et al. [8]. The presence of N-functionalities increases such removals and the removal increase with the increase in the nitrogen content. Li et al. developed high-efficiency and stable visible-light-driven (VLD) photocatalysts and verified that the introduction of CdS in Bi_2MoO_6 enhance the light absorption ability and dramatically boost the separation of charge carriers, leading to excellent performance during toxic antibiotics removal [9].

Summarizing, the present Special Issue reported a detailed preparation of a series of novel catalysts highly efficient in degradation of several pollutants by different AOPs (photocatalysis, catalytic wet air/peroxide oxidation, Fenton based processes, photocatalytic ozonation).

Author Contributions: Conceptualization, O.S.G.P.S., C.A.O., R.P.R. writing—original draft preparation, C.A.O.; writing—review and editing, O.S.G.P.S., C.A.O.; funding acquisition, O.S.G.P.S., C.A.O. All authors have read and agreed to the published version of the manuscript.

Funding: This research was financially supported by InTreat-PTDC/EAM-AMB/31337/2017-POCI-01-0145-FEDER-031337-funded by FEDER funds through COMPETE2020-Programa Operacional Competitividade e Internacionalização (POCI) and by national funds (PIDDAC) through FCT/MCTES and by NORTE-01-0247-FEDER-069836, co-financed by the European Regional Development Fund (ERDF), through the North Portugal Regional Operational Programme (NORTE2020), under the PORTUGAL 2020 Partnership Agreement; LA/P/0045/2020 (ALiCE), UIDB/50020/2020 and UIDP/50020/2020 (LSRE-LCM), funded by national funds through FCT/MCTES (PIDDAC). C.A.O. acknowledges FCT funding under DL57/2016 Transitory Norm Programme. O.S.G.P.S. acknowledges FCT funding under the Scientific Employment Stimulus-Institutional Call CEECINST/00049/2018.

Conflicts of Interest: The authors declare no conflict of interest.

References

1. Rocha, R.; Soares, O.; Órfão, J.; Pereira, M.; Figueiredo, J. Heteroatom (N, S) Co-Doped CNTs in the Phenol Oxidation by Catalytic Wet Air Oxidation. *Catalysts* **2021**, *11*, 578. [CrossRef]
2. Martín-Sómer, M.; Benz, D.; van Ommen, J.; Marugán, J. Multitarget Evaluation of the Photocatalytic Activity of P25-SiO_2 Prepared by Atomic Layer Deposition. *Catalysts* **2020**, *10*, 450. [CrossRef]
3. Sharif, F.; Roberts, E. Electrochemical Oxidation of an Organic Dye Adsorbed on Tin Oxide and Antimony Doped Tin Oxide Graphene Composites. *Catalysts* **2020**, *10*, 263. [CrossRef]
4. Feijoo, S.; González-Rodríguez, J.; Fernández, L.; Vázquez-Vázquez, C.; Feijoo, G.; Moreira, M. Fenton and Photo-Fenton Nanocatalysts Revisited from the Perspective of Life Cycle Assessment. *Catalysts* **2020**, *10*, 23. [CrossRef]
5. Wang, Y.; Zhu, X.; Feng, D.; Hodge, A.; Hu, L.; Lü, J.; Li, J. Biochar-Supported FeS/Fe_3O_4 Composite for Catalyzed Fenton-Type Degradation of Ciprofloxacin. *Catalysts* **2019**, *9*, 1062. [CrossRef]
6. Santos Silva, A.; Seitovna Kalmakhanova, M.; Kabykenovna Massalimova, B.G.; Sgorlon, J.; Jose Luis, D.T.; Gomes, H. Wet Peroxide Oxidation of Paracetamol Using Acid Activated and Fe/Co-Pillared Clay Catalysts Prepared from Natural Clays. *Catalysts* **2019**, *9*, 705. [CrossRef]
7. Orge, C.; Soares, O.; Ramalho, P.; Pereira, M.; Faria, J. Magnetic Nanoparticles for Photocatalytic Ozonation of Organic Pollutants. *Catalysts* **2019**, *9*, 703. [CrossRef]
8. Soares, O.; Rodrigues, C.; Madeira, L.; Pereira, M. Heterogeneous Fenton-Like Degradation of p-Nitrophenol over Tailored Carbon-Based Materials. *Catalysts* **2019**, *9*, 258. [CrossRef]
9. Li, S.; Liu, Y.; Long, Y.; Mo, L.; Zhang, H.; Liu, J. Facile Synthesis of Bi_2MoO_6 Microspheres Decorated by CdS Nanoparticles with Efficient Photocatalytic Removal of Levfloxacin Antibiotic. *Catalysts* **2018**, *8*, 477. [CrossRef]

MDPI

Article

Heteroatom (N, S) Co-Doped CNTs in the Phenol Oxidation by Catalytic Wet Air Oxidation

Raquel P. Rocha *, Olívia Salomé G. P. Soares , José J. M. Órfão, Manuel Fernando R. Pereira and José L. Figueiredo

Laboratory of Separation and Reaction Engineering-Laboratory of Catalysis and Materials (LSRE-LCM), Department of Chemical Engineering, Faculty of Engineering, University of Porto, Rua Dr. Roberto Frias, 4200-465 Porto, Portugal; salome.soares@fe.up.pt (O.S.G.P.S.); jjmo@fe.up.pt (J.J.M.Ó.); fpereira@fe.up.pt (M.F.R.P.); jlfig@fe.up.pt (J.L.F.)

* Correspondence: rprocha@fe.up.pt; Tel.: +351-22-041-4874

Abstract: The N, S-co-doping of commercial carbon nanotubes (CNTs) was performed by a solvent-free mechanothermal approach using thiourea. CNTs were mixed with the N, S-dual precursor in a ball-milling apparatus, and further thermally treated under inert atmosphere between 600 and 1000 °C. The influence of the temperature applied during the thermal procedure was investigated. Textural properties of the materials were not significantly affected either by the mechanical step or by the heating phase. Concerning surface chemistry, the developed methodology allowed the incorporation of N (up to 1.43%) and S (up to 1.3%), distributed by pyridinic (N6), pyrrolic (N5), and quaternary N (NQ) groups, and C–S–, C–S–O, and sulphate functionalities. Catalytic activities of the N, S-doped CNTs were evaluated for the catalytic wet air oxidation (CWAO) of phenol in a batch mode. Although the samples revealed a similar catalytic activity for phenol degradation, a higher total organic carbon removal (60%) was observed using the sample thermally treated at 900 °C. The improved catalytic activity of this sample was attributed to the presence of N6, NQ, and thiophenic groups. This sample was further tested in the oxidation of phenol under a continuous mode, at around 30% of conversion being achieved in the steady-state.

Keywords: metal-free carbon catalysts; N, S-co-doping; carbon nanotubes; catalytic wet air oxidation

Citation: Rocha, R.P.; Soares, O.S.G.P.; Órfão, J.J.M.; Pereira, M.F.R.; Figueiredo, J.L. Heteroatom (N, S) Co-Doped CNTs in the Phenol Oxidation by Catalytic Wet Air Oxidation. *Catalysts* **2021**, *11*, 578. https://doi.org/10.3390/catal11050578

Academic Editor: Francisco Javier Rivas Toledo

Received: 9 April 2021
Accepted: 26 April 2021
Published: 30 April 2021

Publisher's Note: MDPI stays neutral with regard to jurisdictional claims in published maps and institutional affiliations.

1. Introduction

Wet air oxidation (WAO) is an alternative to conventional technologies for the treatment of organic pollutants in wastewaters, especially from highly pollutant industries such as pulp and paper industries, petrochemicals, and wine distilleries [1,2]. The process is suitable to treat organic compounds that show some resistance to conventional treatment technologies or to treat effluents presenting concentrations too high to be treated by biological processes or too low for incineration. It is classified as an advanced oxidation process (AOP), since it is based on the production and use of strongly reactive radicals to mineralise the organic compounds into CO_2 and H_2O or into easily biodegradable by-products. In this case, the strongly oxidizing radicals are formed by the use of pure oxygen or air. To ensure the solubility of oxygen and to promote fast mineralisation and degradation rates, high temperatures (200–320 °C) and pressures (20–200 bar) are required [3–5]. In the last decades, efforts have been made to decrease such severe operating conditions by the addition of homogeneous or heterogeneous catalysts, mostly based on the use of noble metals and metal oxides [6–13]. However, the scarcity of these materials, and the frequent leaching and deactivation phenomena, have prevented the design of a clean catalytic wet air oxidation (CWAO) process [7,14]. During some of these studies, carbon materials were tested as supports for these metals and oxides, and it was found that the carbon material could act as a catalyst on its own, and not only as a support. This opened the opportunity for carbon materials as an alternative metal-free catalyst for CWAO [15–25]. Along the years, mesoporous carbon xerogels, carbon fibers and foams, carbon blacks, carbon nanotubes (CNTs) and nanofibers (CNFs), and also graphene-based materials were investigated in CWAO [26–33]. The outstanding ability to tune the textural and chemical properties of carbon materials make them potential catalyst candidates for a sustainable technology. To improve the performance of carbon materials in CWAO, several authors have taken advantage of the unsaturated carbon atoms at the edges of the graphene layers and at basal plane defects to form various types of surface functional groups containing O, N, S, B or P [34–40]. However, almost all of these studies were dedicated to studying each isolated heteroatom (N, S, B, P), while synergic benefits of using two or more heteroatoms have been reported in the literature [41–43], at least in the field of renewable energy, where metal-free carbon materials have also been widely studied [44–48]. As far as we know, the combination of two heteroatoms, particularly N and S, was never investigated in previous works as metal-free catalysts for CWAO. Exploiting our expertise in developing novel catalysts using fine surface functionalisation of carbon materials by easy-to-handle solvent-free methodologies [34,37,38,49], a systematic study was performed to assess the influence of surface N and S heteroatoms in CWAO. Nitrogen and sulphur have been selected since both can act as electron donors, and show different changes in the electronic density of states [50] that affect π electrons in the carbon lattice. Thiourea was selected as a precursor for the N, S-doping of CNTs.

2. Results and Discussion

2.1. Materials Characterisation

Table 1 summarises the textural properties of the prepared samples. The textural properties were determined from the N_2 adsorption isotherms, with the available surface area of the samples being determined according to the standard Brunauer, Emmett, and Teller method (S_{BET}). The mechanical and thermal treatment promoted an increase of S_{BET} from 291 to 350 $m^2\ g^{-1}$, when the pristine CNT and CNT@TU600 are compared, respectively. The slight increase may be due to the opening of the CNT closed tips during ball milling, commonly observed when CNTs and powder precursors are mixed under ball milling [37]. The thermal treatments applied at higher temperatures did not promote additional changes in S_{BET} (differences smaller than 24 $m^2\ g^{-1}$, which is close to the estimated experimental error of 20 $m^2\ g^{-1}$). The total pore volume (V_p, experimental error $\pm$0.005 $cm^3\ g^{-1}$) of the CNT@TU600 sample decreases by almost half of the pristine CNT sample (determined from the N_2 uptake at p/p_0 = 0.95), but after that slightly increases for the samples heat treated at 700 °C. For the samples treated at the other temperatures, V_p marginally increases. While in the CNT@TU600 sample a higher agglomeration of the material may cause the abrupt V_p decrease, the increase of the temperature to higher temperatures contributes to the recovery of free space in the CNT bundles.

Table 1. Textural characterisation of N, S-co-doped CNT samples.

	Textural Properties N_2 Adsorption	
Sample	**S_{BET} ($m^2\ g^{-1}$)**	**V_P ($cm^3\ g^{-1}$)**
Pristine CNT	291	1.103
CNT@TU600	350	0.578
CNT@TU700	371	0.690
CNT@TU800	374	0.696
CNT@TU900	362	0.692
CNT@TU1000	363	0.712

The samples' bulk and surface compositions were determined by elemental (C, N, H, S, and O contents) and XPS analyses, respectively (Table 2). N- and S-containing groups were incorporated onto the carbon surface by the mechanical mixture of the CNTs with thiourea during ball milling. With a carbon content of around 95% by elemental analysis, the samples revealed that the continuous temperature increase in the thermal treatment promotes a decrease in the N content of the samples between 1.43 and 0.02%. A constant decrease is also observed concerning the O-content. However, a non-regular trend was found regarding S, with the S-content varying from 0.6 up to 1.3%. Regarding the XPS results (Table 2), a similar bulk/surface composition was noticed.

Table 2. Evolution of bulk and surface composition of N, S-co-doped CNT samples, determined by elemental and XPS analyses, respectively, during the thermal treatment.

	Elemental Analysis				XPS			
Sample	C	N	S	O	C	N	S	O
Pristine CNT	98.9	0.15	0.0	0.8	99.2	n.d.	n.d.	0.8
CNT@TU600	93.7	1.43	0.6	2.4	94.2	1.2	0.3	3.6
CNT@TU700	94.9	1.09	1.3	1.7	95.6	0.8	0.4	2.4
CNT@TU800	95.6	0.81	0.8	1.5	97.1	0.9	0.8	1.2
CNT@TU900	96.3	0.70	0.7	1.3	98.2	0.6	0.3	0.9
CNT@TU1000	95.7	0.02	1.3	1.1	97.9	0.2	0.8	1.1

N.d.: Not detected.

The nature of the N, S-functionalities incorporated on the CNTs was investigated using the N1s and S2p XPS spectra of the samples (Figure 1). The N1s and S2p spectra revealed asymmetric peak shapes, suggesting overlapping of individual peaks. Curve-fittings were performed taking into account the nature N and S-bonds present in the N, S-precursor (thiourea), the possible formed bonds as a function of the temperature of the thermal treatment, in agreement with the literature reported characteristics of binding energies [37,48,51].

For the samples treated at the lowest temperatures (600, 700, and 800 °C), the N1s spectra were deconvoluted into three peaks assuming the presence of quaternary nitrogen (NQ), pyrrolic (N5), and pyridinic (N6) structures. The corresponding binding energies and percentages of the N-functionalities are presented in Table 3. These species are typical of carbonaceous surfaces with N-containing precursors treated at temperatures above 550 °C. Despite the reduction of the N-content with the temperature of the thermal treatment, the N6 structure was always the most abundant, the disappearance of the N5 group being observed as the temperature increased.

Table 3. Relative peak contents and positions obtained by the N1s and S2p spectra fitting of the N, S-co-doped CNT samples.

	Peak #1 (N6)		Peak #2 (N5)		Peak #6 (NQ)		Peak #1 (*C*–*S*)		Peak #2 (*S*–*O*)		Peak #3 (Sulphate)	
Sample	B.E. (eV)	% Rel.	B.E. (eV)	% Rel.	B.E. (eV)	% Rel.	B.E. (eV)	% Rel.	B.E. (eV)	% Rel.	B.E. (eV)	% Rel.
CNT@TU600	398.2	50.0	399.7	39.4	401.2	10.6	163.7	53.2	164.5	18.8	168.7	28.0
CNT@TU700	398.2	65.0	399.8	24.8	400.6	10.2	163.4	59.9	164.0	24.8	167.7	15.3
CNT@TU800	398.3	55.4	399.8	18.7	400.6	25.9	163.6	51.7	164.2	23.4	168.7	24.9
CNT@TU900	398.2	67.0	—	—	400.1	33.0	163.8	100.0	—	—	—	—
CNT@TU1000	398.3	57.8	—	—	400.3	42.2	163.7	100.0	—	—	—	—

In the S2p spectra of the CNT@TU600/700/800 samples, three components were identified: C–S bonds around 163.6 ± 0.2 eV, C–S–O species at 164.3 ± 0.2 eV, and sulphate species at 168.2 ± 0.5 eV. It was assumed that the C–S and C–S–O species have S2p3/2 and 2p1/2 doublets separated by 1.18 eV and with an intensity ratio of 2:1 [48,51,52]. The fitted peak positions and relative areas are shown in Table 3. Since the precursor used does not contain oxygen in its composition, these sulphate species may have been formed by the oxidation of sulphur during air exposure [52,53]. However, at temperatures higher than 900 °C, this sulphate completely disappears. After the thermal treatments at 900 °C, the shape and the peaks identified in the S2p spectra drastically changed, and only C–S bonds have been identified. This suggests that C–S–O species and sulphates are thermally degraded beyond 900 °C.

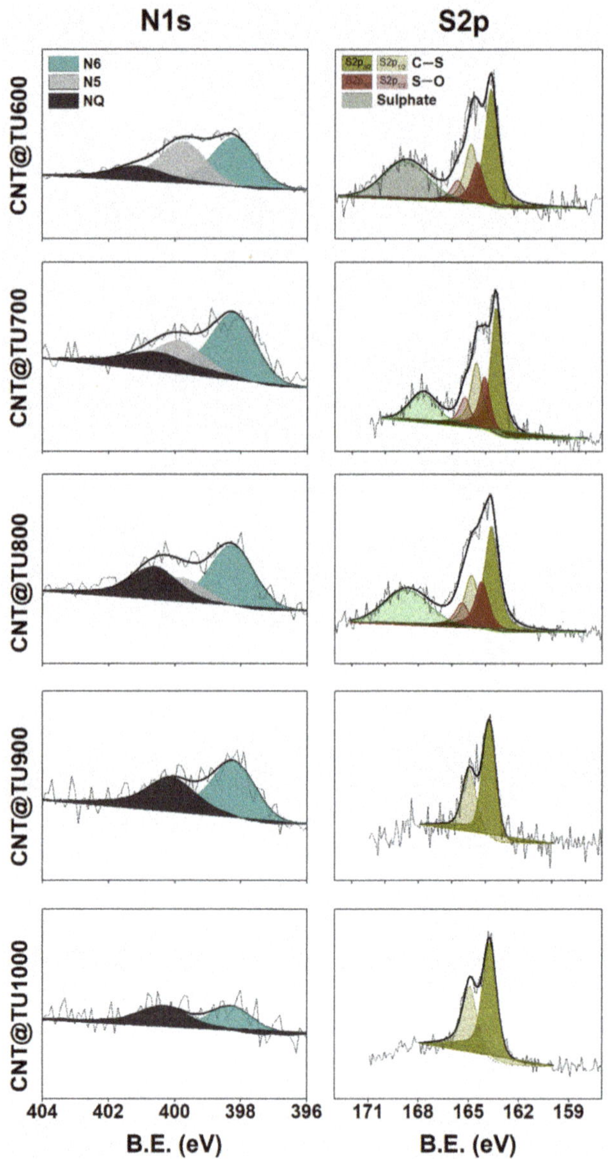

Figure 1. The N1s and S2p XPS spectra for the N, S-co-doped CNT samples.

2.2. Catalytic Activity in CWAO

The catalytic activity of the N, S-co-doped CNTs was preliminary assessed using oxalic acid as a representative of low molecular weight carboxylic acids. This compound has been extensively used in our previous works leading to the development of metal-free carbon materials for CWAO. Under the experimental conditions described elsewhere [34], an extremely fast removal of oxalic acid was observed (total conversion in less than 5 min of reaction using the CNT@TU600 sample, results are not shown). Despite the outstanding catalytic performance over oxalic acid oxidation, such rapid removal will not allow a clear distinction between the catalytic activity of the prepared samples and, for this reason, phenol has been chosen as an alternative model compound. Phenol is a more complex molecule also commonly found in industrial wastes, and extensively studied in the literature as probe species to evaluate the process efficiency. Figure 2a shows the evolution of phenol degradation during the CWAO reaction using the pristine CNT and N, S-CNT samples in a batch mode. In the absence of a catalyst (WAO curve), phenol is not oxidised under the operating conditions employed. Using the pristine CNT sample it is possible to achieve around 50% of phenol degradation after 120 min of reaction. The introduction of the N, S-co-doped CNT catalysts improved the degradation of the aromatic compound to removals between 60 and 70%, after 120 min of reaction, wherein around 50% conversion is obtained in the first 30 min in the presence of the most active samples. However, the TOC removal measured at the end of the catalytic experiments (Figure 2b), revealed a lower removal percentage. While in WAO the difference between the TOC decay and phenol concentration abatement is near 4 percentage points, experiments using the N, S-CNT samples, revealed that those difference may be 4 up to 30 percentage points. The highest TOC conversion (around 60%, corresponding to a phenol removal of 71%) was obtained using the CNT@TU900 sample. Due to the non-microporous nature of the CNTs, as expected, there is no contribution of adsorption to the presented results (results are not shown).

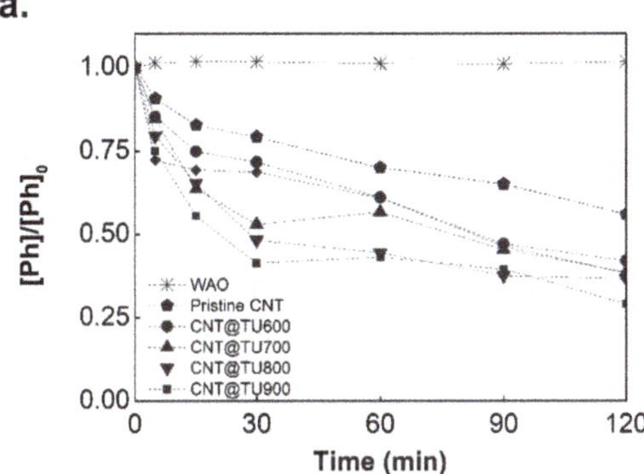

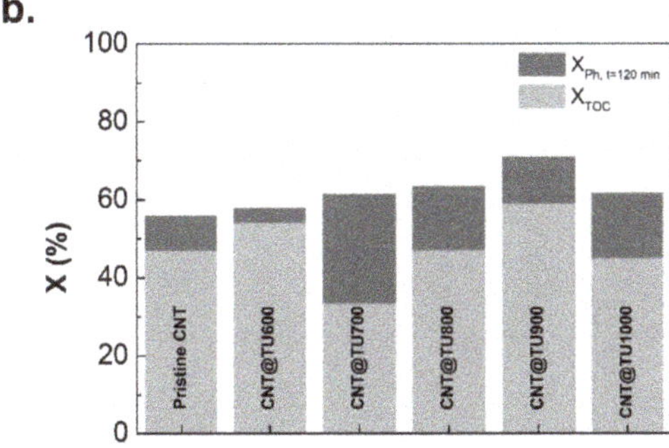

Figure 2. (**a**) Evolution of normalised phenol concentration in a batch mode at 160 °C and 60 bar of total pressure under non-catalytic conditions (WAO) and using the N, S-co-doped CNT samples in CWAO. (**b**) Phenol degradation and TOC removal after 120 min of reaction.

According to the previous works, the increase of the available surface area and the incorporation of N-groups on the CNTs improve the catalytic activity in comparison to the pristine CNTs [34,35,37,38]. In this set of samples, the effect of the surface area is not measurable, since the difference in S_{BET} is smaller than 24 m^2 g^{-1}. Regarding the N, S or O contents, the CNT@TU900 sample does not correspond to the richest sample in any of those elements and, therefore, no correlation between the amounts of N, S or O and the degradation extension is observed. However, among the prepared samples, CNT@TU900 presents a distinct nature of N, S-functionalities. As reported in the materials characterisation section, the thermal treatment at 900 °C revealed distinct N1s and S2p spectra, showing the disappearance of some species. The S2p spectrum of this sample revealed only the presence of C–S species, unlike the samples treated at lower temperatures, where other S-functionalities were also present. While sulfur in the thiophenic (C–S–C) configuration is presumed to induce strain and defects in the carbon matrix, which facilitate a charge localisation for favorable O_2 chemisorption [50], the C–S–O groups present in the samples treated at lower temperatures may decompose forming strongly oxidant sulphate radicals. Although these radicals can be responsible for initiating the degradation of phenol in the liquid phase, they do not allow high levels of mineralisation, suggesting that they can be selective for phenol oxidation but are not able to degrade the various by-products formed. In a previous paper [36], we have observed a similar behavior with $-SO_3H$ groups, promoting low TOC removals in phenol oxidation by CWAO. By other way, the CNT@TU900 sample is also rich in N6 and NQ functionalities, groups that increase the electron density of the surface, enhancing chemisorption and oxygen activation. This explains the better catalytic performance of the CNT@TU900 sample in comparison with the CNT@TU1000 sample, that in addition to not having the S–O and sulphate groups, is also poor in electron donor N-containing groups, resulting in a catalyst with lower catalytic activity.

In order to check the activity and stability of the CNT@TU900 catalyst, a continuous CWAO experiment was carried out until the steady-state was reached (Figure 3). A severe deactivation of the catalyst is observed in the first 10 h, becoming less pronounced after 20 h, with a phenol removal near 27% after 75 h. This represents a slight improvement in the face of the reported result obtained using N-doped CNTs by Santos et al. [54] under similar conditions (a phenol removal near 15% for a similar time). In the previous work, the deactivation phenomenon was also observed and explained by the deposition of phenolic polymers on the surface of the catalyst [55], in which the catalyst is not able to decompose. Although it has not been studied in this work, sometimes it is possible to regenerate the catalyst by carrying out thermal treatments to remove those deposited compounds that block the access to the active sites.

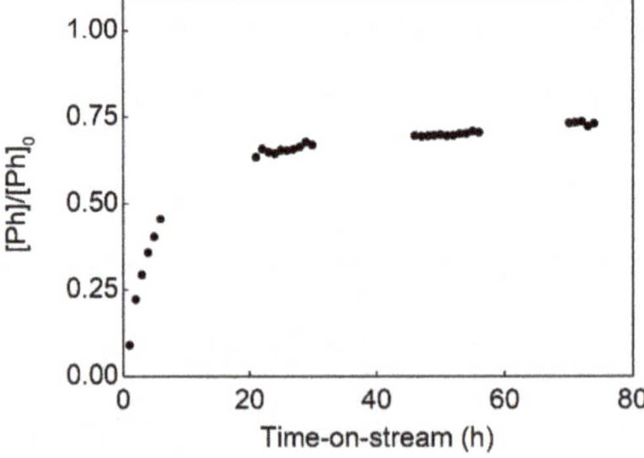

Figure 3. Evolution of normalized phenol concentration in a continuous mode using the CNT@TU900 sample ($[Ph]_0$ = 75 mg L^{-1}; T = 160 °C; P_{O2} = 8 bar; liquid flow rate = 0.25 mL min^{-1}).

3. Experimental

3.1. Materials: Preparation and Characterisation

Commercial multiwalled carbon nanotubes (CNTs) purchased from Nanocyl™ (NC3100 series) were subjected to a mechanothermal treatment to incorporate N, S-heteroatoms onto the graphitic lattice. Thiourea was selected as a precursor for the N, S-co-doping. Briefly, 0.6 g of CNTs were ball milled with thiourea (equivalent to 0.26 g of N). The grinding step was carried out in a Retsch MM200 equipment (Retsch GmbH) using zirconium oxide balls inside the grinding jars without any gas flow. The CNTs and thiourea were mixed and homogenised during 4 h by the radial oscillations of the grinding jars in a horizontal position (15 vibrations s^{-1}), using milling conditions determined in a previous work [37]. After the mechanical mixing, the samples were subjected to a thermal treatment under N_2 (100 cm^3 min^{-1}, at 10 °C min^{-1}) until 600, 700, 800, 900, and 1000 °C, and kept at the final temperature during 1 h. The corresponding samples were designated as CNT@TU, followed by the temperature of the thermal treatment (e.g., CNT@TU600 for the sample thermally treated at 600 °C).

The developed materials were further characterised using the appropriate techniques. Textural properties were based on the N_2 equilibrium adsorption isotherms determined at −196 °C in a Quantachrome NOVA 4200e multi-station apparatus. The surface chemistry was assessed using X-ray photoelectron spectroscopy (XPS). Analyses were performed using a Kratos AXIS Ultra HAS with VISION software for data acquisition and CASAXPS software for data analysis. The chemical composition was evaluated by elemental (CHNS/O) analysis carried out on a vario MICRO cube and a rapid OXY cube analyser from Elemental GmbH. Thermogravimetric analyses (TGA) were carried out using a STA 409 PC/4/H Luxx Netzsch thermal analyser.

3.2. Experimental Procedure

3.2.1. Batch Mode

CWAO experiments in a batch mode were carried in a 160 mL 316-stainless-steel high-pressure batch reactor (Parr Instruments, Moline, IL, Mod. 4564, USA) using phenol as a model pollutant (detailed description elsewhere [39]). In addition, 75 mL of a 75 mg L^{-1} phenol solution and 0.1 g of catalyst were placed inside the reactor. The reactor was purged with nitrogen until the complete removal of oxygen and then pressurised with 5 bar of nitrogen and pre-heated up to 160 °C under continuous stirring (500 rpm). After reaching the desired temperature, pure air was injected until a total pressure of 60 bar inside the reactor was reached. This was considered time zero for the reaction.

3.2.2. Continuous Mode

CWAO experiments in a continuous mode were carried out in a completely automated tubular reactor from Microactivity PID Eng&Tech equipped with a HPLC pump and a high-pressure liquid–gas separator working in down-flow [39]. In addition, 0.2 g of the catalyst were mixed with 1.8 g of carborundum and introduced into the tubular reactor. The system was pre-heated under nitrogen flow (4 bar of pressure) until the desired temperature (160 °C). Then, nitrogen was replaced by pure oxygen, the pressure inside the reactor increased to the desired value (8 bar), and the model pollutant solution (75 mg L^{-1} of phenol) was fed at 0.25 mL min^{-1}.

The operating conditions have been selected taking into account the previous study of Santos et al. [54], where the influence of distinct operating conditions in the same experimental system and using phenol as a model compound, for batch and continuous mode operations, was investigated.

3.3. Analytical Techniques

Liquid aliquots were withdrawn during the reactions and centrifuged (if necessary) for further analysis by high performance liquid chromatography (HPLC) with a Hitachi Elite LAChrom system equipped with a Diode Array Detector (L-2450). The samples were analysed in a Purospher Star RP-18 endcapped column (250 × 4.6 m^2, 5 μm particles) working at room temperature. The mobile phase was a mixture of water and methanol 40/60 (*v*/*v*) with a flow rate of 0.8 mL min^{-1}. Analyses were made using an injection volume of 50 μL. The quantification of phenol was performed at λ = 270 nm. The total organic carbon (TOC) was also determined for the selected experiments in a TOC-L Analyser from Shimadzu by the non-purgable organic carbon (NPOC) method.

4. Conclusions

The introduction of N, S-functional groups onto the carbon surface of commercial CNTs has been successfully carried out using a mechanothermal approach with thiourea as a precursor for the N, S heteroatoms. Without significant damage of the textural properties of the CNTs, quaternary nitrogen (NQ), pyrrolic (N5) and pyridinic (N6) structures, as well as C–S, C–S–O, and sulphate species, were incorporated at different temperatures in the thermal step. Some of these specific groups were found to improve the catalytic activity of the N, S-doped CNT towards phenol oxidation by CWAO, with around 60% of TOC removal being achieved with the CNT@TU900 sample. The enhanced phenol degradation and TOC removal were explained by the synergic contribution of the N6 and NQ groups that increase the electron density of the surface, enhancing chemisorption and oxygen activation, and the presence of thiophenic-like groups (C–S–C), responsible for promoting charge localisation for favorable O_2 chemisorption.

Author Contributions: Writing—original draft preparation, investigation, methodology, conceptualization, R.P.R.; writing—reviewing and editing, conceptualization, O.S.G.P.S.; writing—reviewing and editing, conceptualization, J.J.M.Ó.; supervision, conceptualization, funding acquisition, writing—reviewing and editing, M.F.R.P.; supervision, conceptualization, writing—reviewing and editing, resources, J.L.F. All authors have read and agreed to the published version of the manuscript.

Funding: This work was financially supported by Base Funding—UIDB/50020/2020 of the Associate Laboratory LSRE-LCM—funded by national funds through FCT/MCTES (PIDDAC) and by NORTE-01-0145-FEDER-000054 funded by CCDR-N (Norte2020). O.S.G.P.S. acknowledges FCT funding under the Scientific Employment Stimulus—Institutional Call CEECINST/00049/2018.

Data Availability Statement: Data is contained within the article.

Acknowledgments: The authors are indebted to Carlos M. Sá (CEMUP) for assistance with XPS analyses.

Conflicts of Interest: The authors declare no conflict of interest.

References

1. Mishra, V.S.; Mahajani, V.V.; Joshi, J.B. Wet Air Oxidation. *Ind. Eng. Chem. Res.* **1995**, *34*, 2–48. [CrossRef]
2. Mantzavinos, D.; Sahibzada, M.; Livingston, A.G.; Metcalfe, I.S.; Hellgardt, K. Wastewater treatment: Wet air oxidation as a precursor to biological treatment. *Catal. Today* **1999**, *53*, 93–106. [CrossRef]
3. Kolaczkowski, S.T.; Plucinski, P.; Beltran, F.J.; Rivas, F.J.; McLurgh, D.B. Wet air oxidation: A review of process technologies and aspects in reactor design. *Chem. Eng. J.* **1999**, *73*, 143–160. [CrossRef]
4. Luck, F. Wet air oxidation: Past, present and future. *Catal. Today* **1999**, *53*, 81–91. [CrossRef]
5. Debellefontaine, H.; Foussard, J.N. Wet air oxidation for the treatment of industrial wastes. Chemical aspects, reactor design and industrial applications in Europe. *Waste Manag.* **2000**, *20*, 15–25. [CrossRef]
6. Bhargava, S.K.; Tardio, J.; Prasad, J.; Föger, K.; Akolekar, D.B.; Grocott, S.C. Wet oxidation and catalytic wet oxidation. *Ind. Eng. Chem. Res.* **2006**, *45*, 1221–1258. [CrossRef]
7. Levec, J.; Pintar, A. Catalytic wet-air oxidation processes: A review. *Catal. Today* **2007**, *124*, 172–184. [CrossRef]
8. Gallezot, P.; Laurain, N.; Isnard, P. Catalytic wet-air oxidation of carboxylic acids on carbon-supported platinum catalysts. *Appl. Catal. B Environ.* **1996**, *9*, L11–L17. [CrossRef]
9. Roy, S.; Saroha, A.K. Ceria promoted [gamma]-Al2O3 supported platinum catalyst for catalytic wet air oxidation of oxalic acid: Kinetics and catalyst deactivation. *RSC Adv.* **2014**, *4*, 56838–56847. [CrossRef]
10. Pruden, B.B.; Le, H. Wet air oxidation of soluble components in waste water. *Can. J. Chem. Eng.* **1976**, *54*, 319–325. [CrossRef]
11. Martín-Hernández, M.; Carrera, J.; Suárez-Ojeda, M.E.; Besson, M.; Descorme, C. Catalytic wet air oxidation of a high strength p-nitrophenol wastewater over Ru and Pt catalysts: Influence of the reaction conditions on biodegradability enhancement. *Appl. Catal. B Environ.* **2012**, *123–124*, 141–150. [CrossRef]

12. Stüber, F.; Font, J.; Fortuny, A.; Bengoa, C.; Eftaxias, A.; Fabregat, A. Carbon materials and catalytic wet air oxidation of organic pollutants in wastewater. *Top. Catal.* **2005**, *33*, 3–50. [CrossRef]
13. Aguilar, C.; García, R.; Soto-Garrido, G.; Arriagada, R. Catalytic wet air oxidation of aqueous ammonia with activated carbon. *Appl. Catal. B Environ.* **2003**, *46*, 229–237. [CrossRef]
14. Cybulski, A. Catalytic Wet Air Oxidation: Are Monolithic Catalysts and Reactors Feasible? *Ind. Eng. Chem. Res.* **2007**, *46*, 4007–4033. [CrossRef]
15. Shende, R.V.; Levec, J. Wet Oxidation Kinetics of Refractory Low Molecular Mass Carboxylic Acids. *Ind. Eng. Chem. Res.* **1999**, *38*, 3830–3837. [CrossRef]
16. Quintanilla, A.; Casas, J.A.; Mohedano, A.F.; Rodríguez, J.J. Reaction pathway of the catalytic wet air oxidation of phenol with a Fe/activated carbon catalyst. *Appl. Catal. B Environ.* **2006**, *67*, 206–216. [CrossRef]
17. Quintanilla, A.; Casas, J.A.; Zazo, J.A.; Mohedano, A.F.; Rodríguez, J.J. Wet air oxidation of phenol at mild conditions with a Fe/activated carbon catalyst. *Appl. Catal. B Environ.* **2006**, *62*, 115–120. [CrossRef]
18. Quintanilla, A.; Casas, J.A.; Rodríguez, J.J. Catalytic wet air oxidation of phenol with modified activated carbons and Fe/activated carbon catalysts. *Appl. Catal. B Environ.* **2007**, *76*, 135–145. [CrossRef]
19. Quintanilla, A.; Casas, J.A.; Rodriguez, J.J.; Kreutzer, M.T.; Kapteijn, F.; Moulijn, J.A. Kinetics of the wet oxidation of phenol over an Fe/activated carbon catalyst. *Int. J. Chem. React. Eng.* **2007**, *5*. [CrossRef]
20. Quintanilla, A.; Fraile, A.F.; Casas, J.A.; Rodríguez, J.J. Phenol oxidation by a sequential CWPO-CWAO treatment with a Fe/AC catalyst. *J. Hazard. Mater.* **2007**, *146*, 582–588. [CrossRef]
21. Quintanilla, A.; Menéndez, N.; Tornero, J.; Casas, J.A.; Rodríguez, J.J. Surface modification of carbon-supported iron catalyst during the wet air oxidation of phenol: Influence on activity, selectivity and stability. *Appl. Catal. B Environ.* **2008**, *81*, 105–114. [CrossRef]
22. Quintanilla, A.; Dominguez, C.M.; Casas, J.A.; Rodriguez, J.J. Emerging catalysts for wet air oxidation process. In *Focus on Catalysis Research: New Developments*; Nova Science Publishers, Inc.: New York, NY, USA, 2012; pp. 237–259.
23. Suarez-Ojeda, M.E.; Stüber, F.; Fortuny, A.; Fabregat, A.; Carrera, J.; Font, J. Catalytic wet air oxidation of substituted phenols using activated carbon as catalyst. *Appl. Catal. B Environ.* **2005**, *58*, 105–114. [CrossRef]
24. Suárez-Ojeda, M.E.; Fabregat, A.; Stüber, F.; Fortuny, A.; Carrera, J.; Font, J. Catalytic wet air oxidation of substituted phenols: Temperature and pressure effect on the pollutant removal, the catalyst preservation and the biodegradability enhancement. *Chem. Eng. J.* **2007**, *132*, 105–115. [CrossRef]
25. Suárez-Ojeda, M.E.; Kim, J.; Carrera, J.; Metcalfe, I.S.; Font, J. Catalytic and non-catalytic wet air oxidation of sodium dodecylbenzene sulfonate: Kinetics and biodegradability enhancement. *J. Hazard. Mater.* **2007**, *144*, 655–662. [CrossRef]
26. Rocha, R.P.; Pereira, M.F.R.; Figueiredo, J.L. Metal-free carbon materials as catalysts for wet air oxidation. *Catal. Today* **2019**. [CrossRef]
27. Eftaxias, A.; Font, J.; Fortuny, A.; Fabregat, A.; Stüber, F. Kinetics of phenol oxidation in a trickle bed reactor over active carbon catalyst. *J. Chem. Technol. Biotechnol.* **2005**, *80*, 677–687. [CrossRef]
28. Eftaxias, A.; Font, J.; Fortuny, A.; Fabregat, A.; Stüber, F. Catalytic wet air oxidation of phenol over active carbon catalyst: Global kinetic modelling using simulated annealing. *Appl. Catal. B Environ.* **2006**, *67*, 12–23. [CrossRef]
29. Santiago, M.; Stüber, F.; Fortuny, A.; Fabregat, A.; Font, J. Modified activated carbons for catalytic wet air oxidation of phenol. *Carbon* **2005**, *43*, 2134–2145. [CrossRef]
30. Cordero, T.; Rodríguez-Mirasol, J.; Bedia, J.; Gomis, S.; Yustos, P.; García-Ochoa, F.; Santos, A. Activated carbon as catalyst in wet oxidation of phenol: Effect of the oxidation reaction on the catalyst properties and stability. *Appl. Catal. B Environ.* **2008**, *81*, 122–131. [CrossRef]
31. Fortuny, A.; Font, J.; Fabregat, A. Wet air oxidation of phenol using active carbon as catalyst. *Appl. Catal. B Environ.* **1998**, *19*, 165–173. [CrossRef]
32. Sousa, J.P.S.; Silva, A.M.T.; Pereira, M.F.R.; Figueiredo, J.L. Wet Air Oxidation of Aniline Using Carbon Foams and Fibers Enriched with Nitrogen. *Sep. Sci. Technol.* **2010**, *45*, 1546–1554. [CrossRef]
33. Tukač, V.; Hanika, J. Catalytic effect of active carbon black Chezacarb in wet oxidation of phenol. *Collect. Czechoslov. Chem. Commun.* **1996**, *61*, 1010–1017. [CrossRef]
34. Soares, O.S.G.P.; Rocha, R.P.; Órfão, J.J.M.; Pereira, M.F.R.; Figueiredo, J.L. Mechanothermal Approach for N-, S-, P-, and B-Doping of Carbon Nanotubes: Methodology and Catalytic Performance in Wet Air Oxidation. *C—J. Carbon Res.* **2019**, *5*, 30. [CrossRef]
35. Rocha, R.P.; Sousa, J.P.S.; Silva, A.M.T.; Pereira, M.F.R.; Figueiredo, J.L. Catalytic activity and stability of multiwalled carbon nanotubes in catalytic wet air oxidation of oxalic acid: The role of the basic nature induced by the surface chemistry. *Appl. Catal. B Environ.* **2011**, *104*, 330–336. [CrossRef]
36. Rocha, R.P.; Silva, A.M.T.; Romero, S.M.M.; Pereira, M.F.R.; Figueiredo, J.L. The role of O- and S-containing surface groups on carbon nanotubes for the elimination of organic pollutants by catalytic wet air oxidation. *Appl. Catal. B Environ.* **2014**, *147*, 314–321. [CrossRef]
37. Soares, O.S.G.P.; Rocha, R.P.; Gonçalves, A.G.; Figueiredo, J.L.; Órfão, J.J.M.; Pereira, M.F.R. Easy method to prepare N-doped carbon nanotubes by ball milling. *Carbon* **2015**, *91*, 114–121. [CrossRef]

38. Soares, O.S.G.P.; Rocha, R.P.; Gonçalves, A.G.; Figueiredo, J.L.; Órfão, J.J.M.; Pereira, M.F.R. Highly active N-doped carbon nanotubes prepared by an easy ball milling method for advanced oxidation processes. *Appl. Catal. B Environ.* **2016**, *192*, 296–303. [CrossRef]
39. Rocha, R.P.; Santos, D.F.M.; Soares, O.S.M.P.; Silva, A.M.T.; Pereira, M.F.R.; Figueiredo, J.L. Metal-Free Catalytic Wet Oxidation: From Powder to Structured Catalyst Using N-Doped Carbon Nanotubes. *Top. Catal.* **2018**, *61*, 1957–1966. [CrossRef]
40. Santos, D.F.M.; Soares, O.S.G.P.; Silva, A.M.T.; Figueiredo, J.L.; Pereira, M.F.R. Degradation and mineralization of oxalic acid using catalytic wet oxidation over carbon coated ceramic monoliths. *J. Environ. Chem. Eng.* **2021**, *9*, 105369. [CrossRef]
41. Li, X.; Yang, S.X.; Zhu, W.P.; Wang, J.B.; Wang, L. Catalytic wet air oxidation of phenol and aniline over multi-walled carbon nanotubes. *Huanjing Kexue/Environ. Sci.* **2008**, *29*, 2522–2528.
42. Yang, S.; Wang, X.; Yang, H.; Sun, Y.; Liu, Y. Influence of the different oxidation treatment on the performance of multi-walled carbon nanotubes in the catalytic wet air oxidation of phenol. *J. Hazard. Mater.* **2012**, *233–234*, 18–24. [CrossRef]
43. Yang, S.; Sun, Y.; Yang, H.; Wan, J. Catalytic wet air oxidation of phenol, nitrobenzene and aniline over the multi-walled carbon nanotubes (MWCNTs) as catalysts. *Front. Environ. Sci. Eng.* **2014**. [CrossRef]
44. He, W.; Xue, P.; Du, H.; Xu, L.; Pang, M.; Gao, X.; Yu, J.; Zhang, Z.; Huang, T. A facile method prepared nitrogen and boron doped carbon nano-tube based catalysts for oxygen reduction. *Int. J. Hydrogen Energy* **2017**, *42*, 4123–4132. [CrossRef]
45. Huang, X.; Wang, Q.; Jiang, D.; Huang, Y. Facile synthesis of B, N co-doped three-dimensional porous graphitic carbon toward oxygen reduction reaction and oxygen evolution reaction. *Catal. Commun.* **2017**, *100*, 89–92. [CrossRef]
46. Liu, Y.; Zhang, Y.; Cheng, K.; Quan, X.; Fan, X.; Su, Y.; Chen, S.; Zhao, H.; Zhang, Y.; Yu, H.; et al. Selective Electrochemical Reduction of Carbon Dioxide to Ethanol on a Boron- and Nitrogen-Co-doped Nanodiamond. *Angew. Chem. Int. Ed.* **2017**, *56*, 15607–15611. [CrossRef]
47. Zeng, L.; Shi, J.; Luo, J.; Chen, H. Silver sulfide anchored on reduced graphene oxide as a high -performance catalyst for CO2 electroreduction. *J. Power Sources* **2018**, *398*, 83–90. [CrossRef]
48. Yuan, Q.; Ma, Z.; Chen, J.; Huang, Z.; Fang, Z.; Zhang, P.; Lin, Z.; Cui, J. N, S-Codoped Activated Carbon Material with Ultra-High Surface Area for High-Performance Supercapacitors. *Polymers* **2020**, *12*, 1982. [CrossRef]
49. Rocha, I.M.; Soares, O.S.G.P.; Fernandes, D.M.; Freire, C.; Figueiredo, J.L.; Pereira, M.F.R. N-doped Carbon Nanotubes for the Oxygen Reduction Reaction in Alkaline Medium: Synergistic Relationship between Pyridinic and Quaternary Nitrogen. *Chem. Sel.* **2016**, *1*, 2522–2530. [CrossRef]
50. Kiciński, W.; Szala, M.; Bystrzejewski, M. Sulfur-doped porous carbons: Synthesis and applications. *Carbon* **2014**, *68*, 1–32. [CrossRef]
51. Zhou, G.; Yin, L.-C.; Wang, D.-W.; Li, L.; Pei, S.; Gentle, I.R.; Li, F.; Cheng, H.-M. Fibrous Hybrid of Graphene and Sulfur Nanocrystals for High-Performance Lithium–Sulfur Batteries. *ACS Nano* **2013**, *7*, 5367–5375. [CrossRef]
52. Zheng, X.; Wu, J.; Cao, X.; Abbott, J.; Jin, C.; Wang, H.; Strasser, P.; Yang, R.; Chen, X.; Wu, G. N-, P-, and S-doped graphene-like carbon catalysts derived from onium salts with enhanced oxygen chemisorption for Zn-air battery cathodes. *Appl. Catal. B Environ.* **2019**, *241*, 442–451. [CrossRef]
53. Schaufuß, A.G.; Nesbitt, H.W.; Kartio, I.; Laajalehto, K.; Bancroft, G.M.; Szargan, R. Incipient oxidation of fractured pyrite surfaces in air. *J. Electron Spectrosc. Relat. Phenom.* **1998**, *96*, 69–82. [CrossRef]
54. Santos, D.F.M.; Soares, O.S.G.P.; Silva, A.M.T.; Figueiredo, J.L.; Pereira, M.F.R. Catalytic wet oxidation of organic compounds over N-doped carbon nanotubes in batch and continuous operation. *Appl. Catal. B Environ.* **2016**, *199*, 361–371. [CrossRef]
55. Delgado, J.J.; Chen, X.; Pérez-Omil, J.A.; Rodríguez-Izquierdo, J.M.; Cauqui, M.A. The effect of reaction conditions on the apparent deactivation of Ce–Zr mixed oxides for the catalytic wet oxidation of phenol. *Catal. Today* **2012**, *180*, 25–33. [CrossRef]

Article

Multitarget Evaluation of the Photocatalytic Activity of P25-SiO_2 Prepared by Atomic Layer Deposition

Miguel Martín-Sómer [1], Dominik Benz [2], J. Ruud van Ommen [2,*] and Javier Marugán [1,*]

1 Department of Chemical and Environmental Technology, Universidad Rey Juan Carlos, C/Tulipán s/n, 28933 Móstoles, Madrid, Spain; miguel.somer@urjc.es
2 Department of Chemical Engineering, Delft University of Technology, Mekelweg 5, 2628 CD Delft, The Netherlands; D.Benz@tudelft.nl
* Correspondence: j.r.vanOmmen@tudelft.nl (J.R.v.O.); javier.marugan@urjc.es (J.M.); Tel.: +34-491-664-7466 (J.M.)

Received: 25 February 2020; Accepted: 21 April 2020; Published: 22 April 2020

Abstract: This work presents the evaluation of the photocatalytic activity of P25 TiO_2 particles, coated with SiO_2, using atomic layer deposition (ALD) for the photocatalytic removal of methylene blue, oxidation of methanol and inactivation of *Escherichia coli* bacteria in water and its comparative evaluation with bare P25 TiO_2. Two different reactor configurations were used, a slurry reactor with the catalyst in suspension, and a structured reactor with the catalyst immobilized in macroporous foams, that enables the long-term operation of the process in continuous mode, without the necessity of separation of the particles. The results show that the incorporation of SiO_2 decreases the efficiency of the photocatalytic oxidation of methanol, whereas a significant improvement in the removal of methylene blue is achieved, and no significant changes are observed in the photocatalytic inactivation of bacteria. Adsorption tests showed that the improvements, observed in the removal of methylene blue by the incorporation of SiO_2, was mainly due to an increase in its adsorption. The improvement in the adsorption step as part of the global photocatalytic process led to a significant increase in its removal efficiency. Similar conclusions were reached for bacterial inactivation where the loss of photocatalytic efficiency, suggested by the methanol oxidation tests, was counteracted with a better adherence of bacteria to the catalyst that improved its elimination. With respect to the use of macroporous foams as support, a reduction in the photocatalytic efficiency is observed, as expected from the decrease in the available surface area. Nevertheless, this lower efficiency can be counteracted by the operational improvement derived from the easy catalyst reuse.

Keywords: atomic layer deposition; water treatment; photocatalysis; TiO_2-SiO_2; immobilized photocatalyst; methylene blue adsorption

1. Introduction

Among the photochemical processes, heterogeneous photocatalysis stands out as one of the most attractive processes for the treatment of effluents with contaminants that cannot be eliminated by conventional water treatment technologies. Photocatalysis has the advantage of having simple operating conditions since it can be carried out at ambient temperature and pressure, using the oxygen from the air as an oxidizing agent.

The commercial material Evonik P25 (before Degussa P25), is by far the most used photocatalyst. It is constituted by a 3:1 ratio between the phases of TiO_2 anatase and rutile [1] and has the advantage of its low toxicity, high active area, stability and low cost. However, one of the main disadvantages of TiO_2 is the necessity of use light in the UV range for its activation. This makes energy consumption the main expense of the process when using artificial light, or makes the process rather inefficient when using sunlight.

Nowadays, new methods are being used to extend the range of wavelengths, in which the TiO_2 is able to absorb light and thus make the ultraviolet and visible spectra usable. Some of the ways to achieve this are sensitization with dyes or coupling of semiconductors [2–5], synthesis of mesoporous TiO_2, the use of different morphologies of TiO_2 at the nanometric level, reduction of agglomeration in TiO_2 nanoparticles or treatments to modify its surface. Another process for achieving improvements in TiO_2 activity consists of doping the TiO_2 with metallic and non-metallic elements. To achieve this, various processes have been successfully used, such as mechanochemistry [6,7], centrifugation coating methods [8], or wet chemistry methods [9]. However, these methods of deposition have some limitations among which are included: Long processing times, low product homogeneity and the necessity to incorporate additional stages for the separation of impurities [10,11].

One technique for doping TiO_2 that avoids the aforementioned problems is atomic layer deposition (ALD) [12]. This technique uses two gas-phase reaction stages directly on the surface of the product (in this case the TiO_2 nanoparticles) to deposit the material layer by layer [13]. First, a first reactant or precursor is passed through the system that adheres to the TiO_2 surface until it reaches saturation. Subsequently, a purge of the excess precursor is carried out, thus, ensuring that only a thin film remains on the surface. Subsequently, the second reactant or precursor that reacts with the first one is dosed. Again, the system is re-purged, ensuring that only precursors remain attached to thin films on the surface. This cycle is repeated the desired number of times to control the thickness reached. Because each stage of precursor exposure saturates the surface with a monomolecular layer of that precursor, the reactions are self-limited, allowing controlling the deposition at the atomic level, giving rise to several very advantageous characteristics, such as excellent formability and uniformity, and thickness control of the film.

In this work, P25 particles were coated with SiO_2 using ALD carried out in a fluidized bed, in order to improve the photocatalysis efficiency. Once the modified P25 (P25-SiO_2) was obtained, its efficiency in bacterial inactivation and in the removal of methylene blue and methanol from the water was compared with that obtained for the use of commercial P25. Additionally, as the synthesis of the catalyst P25-SiO_2 supposes an increase in the total costs of the process, its immobilization in macroporous foams was carried out with the aim of enabling the reuse of the catalyst. The use of these foams has already been shown to lead to comparable efficiencies to those achieved with the use of suspended catalysts [5].

2. Results and Discussion

2.1. Catalyst Characterization

The incorporation of SiO_2 to the TiO_2 material was confirmed by elemental analysis of the materials using ICP-AES (Table 1). The results showed an average percentage of Si of 1.3 wt % (2.78 wt % of SiO_2).

On the other hand, with the aim of achieving a better understanding of the results, the optical properties of the catalysts were studied. Figure 1 shows the extinction coefficients obtained for P25 and P25-SiO_2 catalysts for different wavelengths in the 300 to 450 nm wavelength range. It can be observed that P25-SiO_2 shows an extinction coefficient 25% lower than P25 at 365 nm but maintaining the same trend in relation to wavelength. Although, the extinction coefficient is not a direct measurement of the photonic absorption, because it also includes the scattering of radiation, it can reasonably assumed that both materials would behave similarly from the scattering point of view, and therefore the absorption of the P25-SiO_2 is significantly lower. Consequently, considering that the absorption of radiation is the triggering step of the photocatalytic process, the incorporation of SiO_2 can potentially reduce the reaction rate.

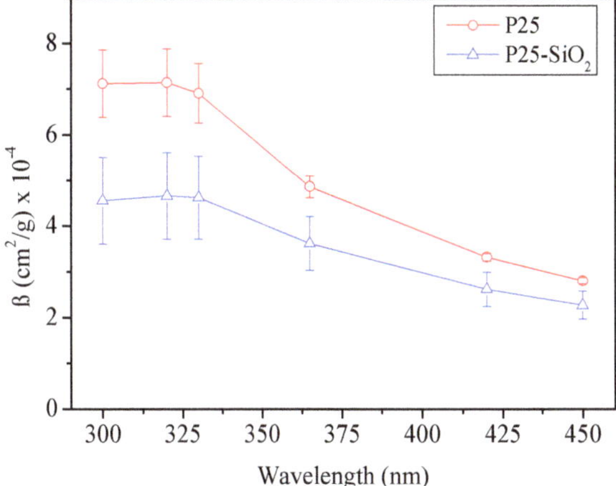

Figure 1. Napierian extinction coefficients for P25 and P25-SiO_2.

On the other hand, in order to check if the addition of SiO_2 produces a variation in the energy that is able to absorb the catalyst, the bandgap was obtained for the two catalysts. Figure 2a shows the diffuse reflectance spectra (DRS) of the three catalysts. Considering the proportionality between F(R) and the absorbance, it is possible to obtain the bandgap by extrapolation from the linear part of the reflectance spectrum represented as $(F(R) \times E)^{1/2}$ vs. E (Figure 2b), as described in bibliography [14].

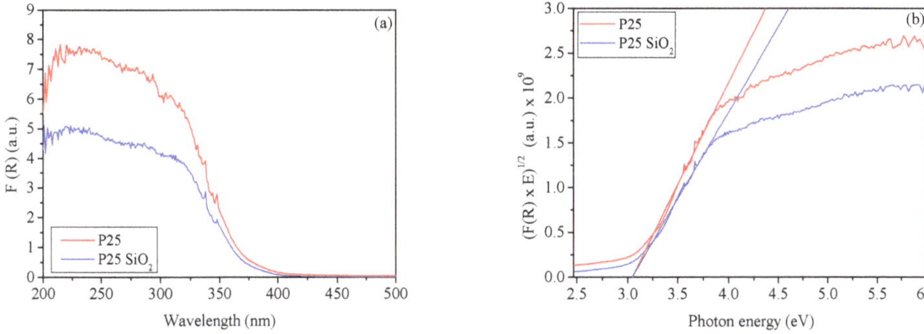

Figure 2. (**a**) DRS spectra; and (**b**) plot of transformed Kubelka–Munk function versus the energy of light for P25 and P25-SiO_2.

As shown in Table 1, bandgap values of 3.05 and 3.04 eV were obtained for P25 and P25-SiO_2, respectively. The values obtained are similar and agree with the bandgap values of the phases of TiO_2 anatase (3.2 eV) and rutile (3 eV), so it can be concluded that the addition of SiO_2 does not produce a significant change of the bandgap of the catalyst.

The textural properties of the catalysts were studied using the N_2 adsorption-desorption isotherms and are shown in Figure 3. How the obtained isotherms can be assimilated to an IUPAC type II isotherm can be observed, indicating the reduced porous character of the materials. The BET surface was calculated (Table 1) and values of 53.3 and 48.7 m^2/g were obtained for P25 and P25-SiO_2, respectively. The slightly smaller surface area of the modified catalyst can be easily explained considering that the addition of SiO_2 reduces the available surface.

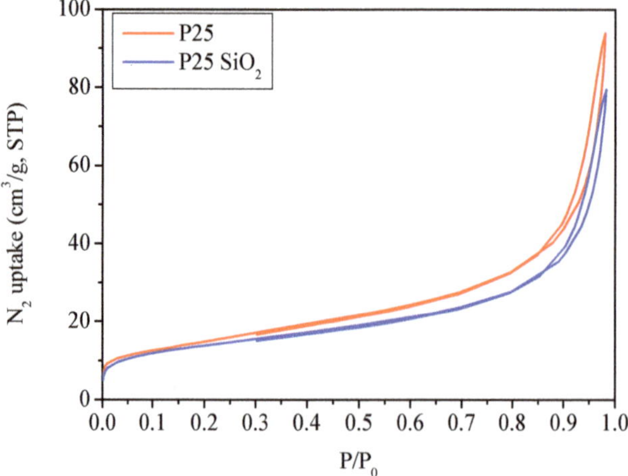

Figure 3. N_2 adsorption-desorption isotherms of P25 and P25-SiO_2.

The zeta potential value obtained for the P25 catalyst at natural pH was 18.1 mV, similar to values found in the literature [15]. However, in the case of the P25-SiO_2 catalyst, due to the negative potential of the silica [16], a final negative potential value of −3.7 mV was obtained. This fact is important, given the different zeta potential of both catalysts can concur in different ways of interacting with the pollutants to be degraded and therefore in different efficiencies in their removal. Moreover, the lower absolute value of the potential of the P25-SiO_2 facilitates the agglomeration of the catalyst particles, reducing the optical density of the suspensions and therefore the light absorption. Table 1 summarizes the main features of for both catalysts.

Table 1. Summary of the main features of P25 and P25-SiO_2.

Feature	TiO_2	SiO_2-TiO_2
SiO_2 (%)	0	2.78
β_{365nm} (cm^2/g)	48,700	36,300
Bandgap (eV)	3.05	3.04
S_{BET} (m^2/g)	53.3	48.7
Zeta Potential (mV)	18.1	−3.7

2.2. Photocatalytic Experiments

Photocatalytic experiments of methanol and methylene blue oxidation under solar irradiation with the catalysis in suspension were carried out in a reactor couple to a compound parabolic collector (CPC) (Figure 4). In the case of the methanol reaction test, no photolytic oxidation takes place [17]. Regarding methylene blue, it could present photolytic removal values of up to 10% [18,19], but this effect would be equivalent for both P25 and P25-SiO_2 materials, and therefore can be neglected for comparative purposes. On the other hand, methylene blue has been extensively studied under irradiation, and shows a certain generation of ROS [20,21] that would increase photocatalytic efficiency. However, again, this behavior is expected to be similar for the two catalysts used, so it can be ignored in this comparative study. Regarding *E. coli* inactivation, a comprehensive study of the analogies and differences between photocatalytic oxidation of chemicals and photocatalytic inactivation of microorganisms, using methylene blue and *E. coli* as target models can be found elsewhere, including the kinetic analysis of the process [22]. The results obtained for methylene blue were fitted to a first-order kinetic model with respect to the accumulated solar incident radiation, usual way of reporting the photocatalytic activity in solar processes [23]. On the other hand, results for methanol oxidation were fitted using the formaldehyde production throughout the reaction following zero-order kinetics as detailed elsewhere [24]. The initial reaction rate for methylene blue was obtained by multiplying the calculated first-order kinetic constant by the initial concentration, whereas in the case of methanol oxidation the initial reaction rate was directly calculated from the fitting to a zero order kinetics of the formaldehyde concentration. The results show an improvement in methylene blue removal when using P25-SiO_2 with respect to the commercial P25. However, in the case of methanol oxidation, the opposite behaviour is observed: The methanol oxidation rate decreases (Figure 4).

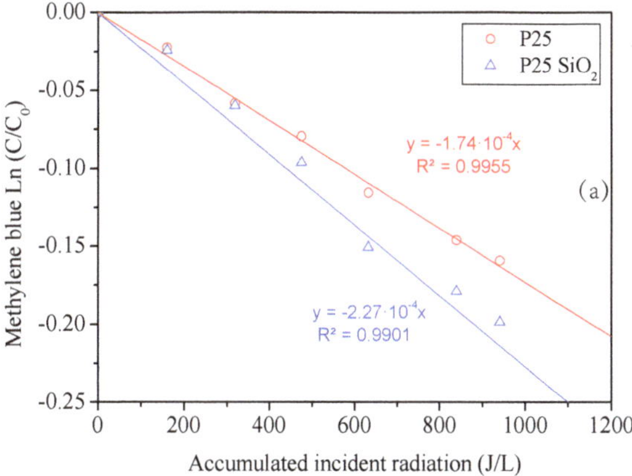

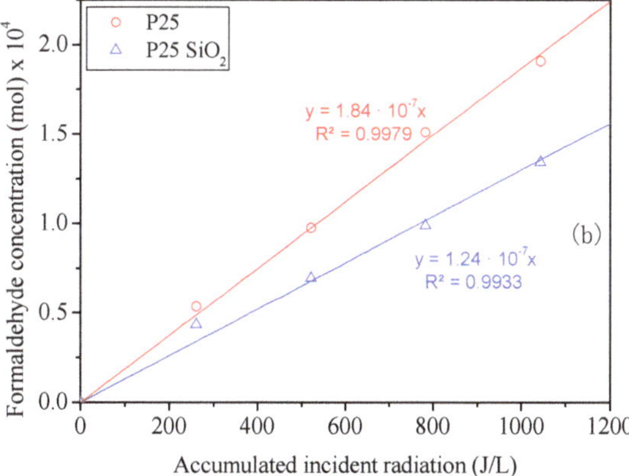

Figure 4. *Cont.*

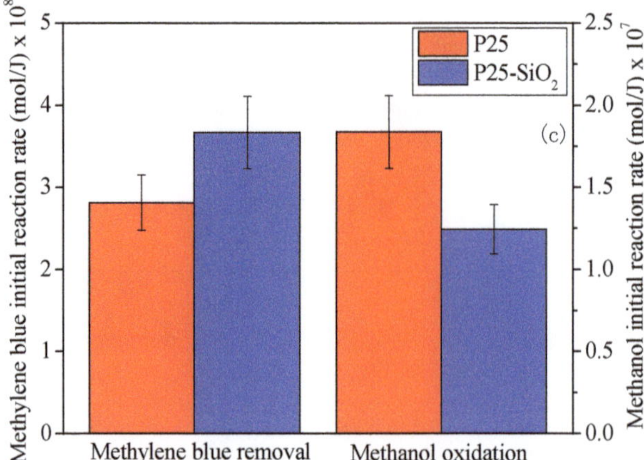

Figure 4. (**a**) First-order kinetic plot of methylene blue removal; (**b**) zero-order kinetic plot of formaldehyde formation; and (**c**) initial reaction rates for both reactions in a CPC reactor under solar irradiation. Error bars calculated from at least three replicate experiments.

Figure 1 shows how the addition of SiO_2 to the P25 catalyst did not modified the spectral response of the catalyst, and only an attenuation in the extinction coefficient was observed (Table 1).In addition, in Figure 2, the non-existence of clear differences in the bandgap values between both catalysts could be confirmed. Therefore, the difference in the results obtained for each catalyst, observed in Figure 4, seems to indicate the existence of different behavior of the catalysts when interacting with the two pollutants, and not the existence of differences in photonic absorption.

In order to find an explanation for the different behavior of the catalysts with each pollutant, similar photocatalytic reactions were carried out in an up-flow reactor, in which photocatalytic foams prepared with both catalysts could also be used. In this case, the light source was a lamp with 40 LEDs of 365 nm. Additionally, the application of the materials to bacterial inactivation processes was also studied to provide new insights. *E. coli* inactivation kinetics were calculated from the logarithmic inactivation profiles using the mechanistic model developed by Marugán et al. [25].

Figure 5a,b shows that results similar to those previously observed when using the CPC reactor under solar irradiation (Figure 4) were obtained, with an improvement in the removal of methylene blue with P25-SiO_2 catalyst, which was not observed in the methanol oxidation. These results confirm that the addition of SiO_2 to P25, do not lead to an amplification in the absorption spectra of the catalyst since the $k_{P25}/k_{P25\text{-}SiO2}$ ratio was similar for both solar and artificial 365 nm irradiation. On the other hand, in the case of bacterial inactivation, it could be observed (Figure 5c) that similar values were obtained for both catalyts.

The results of methanol oxidation tests show a reduction of approximately 25% of the efficiency when using P25-TiO_2 in comparison with the bare P25. This value is similar to the reduction of the extinction coefficient at 365 nm (Table 1), indicating that the decrease in the photocatalytic efficiency is due to the lower photon absorption rate upon incorporation of the SiO_2.

In contrast, in the case of methylene blue removal and E. coli inactivation it seems that the global efficiency is not exclusively limited by the photocatalytic process, suggesting that the adsorption and interaction with the catalyst could play a significant role.

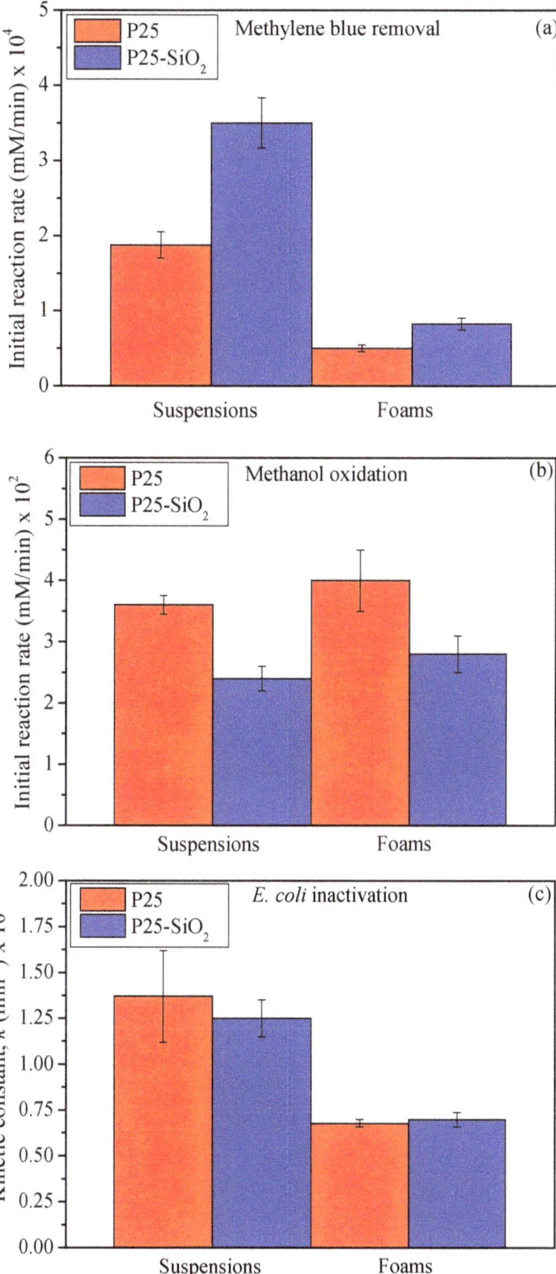

Figure 5. Initial reaction rate for; (**a**) methylene blue removal; and (**b**) methanol oxidation; and (**c**) kinetic constant bacterial inactivation (fitted from the logarithmic inactivation profiles according to Marugán et al. [23]) in an up-flow reactor under artificial 365 nm light. Error bars calculated from at least three replicate experiments.

Some studies have shown the efficiency of SiO_2 to carry out the removal of methylene blue from water by adsorption [26]. Furthermore, it has been demonstrated that the addition of SiO_2 to TiO_2 can improve the adsorption of contaminants [27]. On the other hand, the loss of photocatalytic efficiency that can result from the addition of SiO_2 to the TiO_2 surface, as observed in methanol oxidation, has also been revealed [28].

Taking into account the above, and trying to explain the obtained results, it was decided to carry out methylene blue adsorption experiments. Methylene blue solutions with different concentrations of both P25 and P25-SiO_2 were prepared and the evolution of methylene blue concentration was followed over time. The results shown in Figure 6 indicate that, due to the addition of SiO_2 to the catalyst structure, there is a considerable increase in the adsorption of the methylene blue, which increases the efficiency of its removal from water, arising from the combined effect of adsorption and photocatalytic decomposition. This behavior can be explained if we focus on the results obtained previously in relation to the zeta potential. It was observed that the zeta potential of P25 was positively charged, however, the addition of SiO_2 produced an alteration of the zeta potential that becomes negative (Table 1). Since methylene blue has a positive charge [29], it is easy to understand that it interacts much more with P25-SiO_2. Therefore, being the adsorption capacity of P25-SiO_2 much greater than in the case of P25. Thus, it is possible to explain the greater removal efficiency of methylene blue when using the P25-SiO_2 catalyst due to the synergistic combination of the adsorption and the photocatalytic reaction processes. This synergism accelerates the reaction due to the higher concentration of dye molecules on the catalysts surface, but also accelerates the adsorption due to the consumption of the adsorbed molecules by the reaction, releasing the active sites for the adsorption of new dye molecules. Similar results were obtained by Li et al. [30] in the removal of methylene blue using a P25-graphene composite. They observed that the removal of methylene blue was enhanced by the combination of adsorption and the photocatalytic process. The same conclusions were also reported by other research groups working with different TiO_2 modifications [31,32]. In any case, considering that the reaction time of the photocatalytic experiments is 30 min, and that time is not sufficient to reach the adsorption equilibrium (Figure 6), it cannot be discarded that long-term experiment could present a saturation effect.

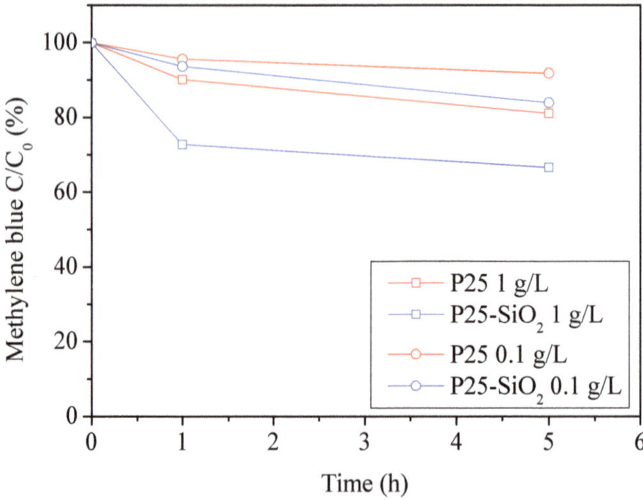

Figure 6. Adsorption of methylene blue along time for different concentrations of P25 and P25-SiO_2.

A similar behaviour may explain the higher efficiency obtained in the *E. coli* inactivation, compared to that expected if only the photocatalytic efficiency is taken into account. Several studies have focused on the adhesion of the bacteria *E. coli* on SiO_2 materials [33,34], showing that the addition of SiO_2 to P25 makes the adhesion of *E. coli* bacteria on the surface of the catalyst became more thermodynamically favorable. Additionally, a higher antibacterial activity was observed for adhered E. coli cells than the suspended cells in aqueous phase, which was explained by the short half-life of reactive oxygen species and slow diffusion in aqueous phase [35].

In relation to the use of photocatalytic foams, Figure 5 shows an inevitable reduction in the removal efficiency of pollutants in comparison with the suspended catalysts, due to a reduction in the surface area available for the reaction. However, the results show very similar removal efficiencies for both catalysts and the loss in activity can be counteracted by the advantages of the use of the immobilized photocatalyst. The deposition of commercial P25 and othersTiO_2 materials on the foam template and its stability through subsequent reaction cycles has been demonstrated in previous work [36].

3. Materials and Methods

3.1. Atomic Layer Deposition

The modification of P25 particles was carried out by adding SiO_2 by ALD. ALD consists of two gas-phase reaction stages directly on the surface of the product to deposit the material layer by layer [37]. For this, a fluidized bed, consisting of a glass column of 25 mm in diameter and 500 mm in length was used. A metal connection was placed in the lower part of the column and in the upper part where the input and output lines were connected. This connection included a distributor plate SIKA-R 20 AX to avoid the exit of the particles to the outside of the reactor and to ensure homogeneous gas distribution at the column entry. The column was placed on a vibrating table Straw PTL 40 / 40-24 (frequency: 35 Hz) to help fluidize the particles. The metal precursor used was $SiCl_4$; it was kept in a stainless steel bubbler (Strem Chemicals, Inc., Newburyport, MA, USA) under an inert gas atmosphere. Pneumatic valves were used to control the flow of gas in and out of the column. The metal precursor and the purge inlet flow contained nitrogen (N_2, grade 5.0), while the oxidizer (precursor 2) inlet flow contained wet N_2. The glass column was heated to 100 °C with an infrared lamp.

To carry out the SiO_2 deposition, 5 g of P25 was introduced into the interior of the column and 8 reaction cycles were carried out. Each reaction cycle consisted of a 30 s dosage of the metal precursor, followed by 5 min of purging. Later, a 3 min dose of oxidant precursor followed by a purge with nitrogen of 8 min. Both the reaction temperatures, the number of cycles and the dosing time of each precursor were previously optimized [38].

3.2. Photocatalytic Experiments

Three different contaminants were tested throughout this study with the aim of having a clear vision of the effects produced by incorporating SiO_2 into commercial P25. On the one hand, chemical oxidation experiments were carried out using both, methylene blue and methanol. The initial concentration of methylene blue and methanol was 0.05 mM, and 100 mM, respectively and the solutions were prepared in deionized water. Photocatalytic degradation of methylene blue was monitored by direct measurement of the absorption at 664 nm in a spectrophotometer while methanol oxidation was followed through the colorimetric determination of the formaldehyde produced throughout the reaction, which is a quantitative oxidation product when methanol is in excess [24]. In addition, bacterial inactivation experiments were carried out. Synthetic wastewater was prepared by adding *E. coli* K12 (CECT 4624, corresponding to ATCC 23631) with an initial concentration of 10^6 CFU/mL. Fresh liquid cultures of *E. coli* were prepared by inoculation in Luria-Bertani (LB) nutrient medium (Miller's LB Broth, Scharlab) and incubation at 37 °C for 24 h under constant stirring on a rotary shaker. The concentration of viable bacteria was quantified throughout the reaction according to the standard serial dilution procedure. Each decimal dilution was spotted 8 times on LB agar plates and incubated at 37 °C before counting after 24 h.

Both the photocatalytic activity of P25 and P25-SiO_2 was tested using two configurations: (i) in suspension and (ii) supported onto three-dimensional (3D) foams (in order to avoid the loss of the photocatalyst). When using the suspensions in all cases, the catalyst concentration was 0.10 g/L and the suspensions were sonicated for 30 min before the reaction.

3.3. Photocatalytic Reactors

A compound parabolic collector (CPC) reactor that operates under sunlight was used to check whether the addition of SiO_2 involved an improvement in the P25 absorption of the solar spectrum. The CPC has two differentiated circuits, in which experiments were carried out simultaneously with P25 and P25-SiO_2 to ensure comparison under exactly the same sunlight conditions. Each circuit has a borosilicate 3.3 Duran®glass tube placed in the focal line of the CPC collector with a length of 380 mm and an inner diameter of 26 mm [39]. The reactor was operated in a closed recirculating circuit driven with a reservoir tank, being the total working volume of 1 L. The experiments were carried in June 2019 at Universidad Rey Juan Carlos facilities in Mostoles, Spain (40.33° N, 3.88° W). The solar irradiance was monitored during the reaction time with a spectrophotometer (Blue Wave, StellarNetInc., Tampa, FL, USA).

An up-flow annular reactor (15 cm long, 3 cm inner diameter and 5 cm outer diameter) was also used. The reactor operated in a closed recirculating circuit driven with a reservoir tank, being the total working volume of 1 L. As illumination source, a 40 LED system (LedEngin Model LZ1-00UV00) with maximum emission peak centered in 365 nm was placed in the axis of the reactor. This reactor allows the operation not only with the catalyst in suspension but also with the photocatalytic foams, as reported elsewhere [36].

3.4. Foams Coating

Macroporous ZnO foams supplied by Insertec SA (2.5 cm height, 3 cm inner diameter, 5 cm outer diameter and a porosity of 10 ppi were coated with P25 and modified P25 (P25-SiO_2) according to the procedure described in a previous work [5]. First, catalyst solutions were prepared in distilled water at a concentration of 10 g/L, and then the foam template was dipped in the solutions and removed after a few seconds. The excess of catalyst was removed by the use of air. Then, the foams were dried for two hours at 100 °C and subsequently calcined for two hours at 500 °C. The amount of catalyst deposited in each foam was measured by weighing difference and the procedure was repeated until reach a total weight in each foam close to 1 g as was previously established as optimum in a previous work where it was also concluded that the foam templates does not show any photocatalytic activity [5]. The photocatalytic experiments for each catalyst were carried out using six foams being the total working height 15 cm.

3.5. Characterization Techniques

The amount of Si deposited in the TiO_2 was measured by inductively coupled plasma atomic emission spectroscopy (ICP-AES) using a Perkin Elmer Optima 4300DV instrument (nebulizer: PE / injector: Al_2O_3). Approximately 30 mg of sample was dissolved in 4.5 mL 30% HCl + 1.5 mL 65% HNO_3 + 0.2 mL 40% HF using a microwave during 60 min. After dissolution the samples were diluted to 50 mL with deionized water and analyzed by ICP-AES.

The specific extinction coefficients (β^*) of the catalysts were obtained using the direct measurement of the transmittance of suspensions with increasing concentration of the material in a UV/vis spectrophotometer (UV-Vis-NIR Varian Cary 500).

Nitrogen adsorption-desorption isotherms were measured using an AutoSorb equipment (Quantachrome Instruments). Degassing of the materials prior to the analysis was carried out by heating at 373 K and applying vacuum until 1×10^{-3} kPa. The surface area was calculated by using the Brunauer–Emmett–Teller (BET) model using the adsorption branch. Zeta potential values were measured using a NanoPlus DLS Zeta Potential equipment to determine the electrostatic interactions among catalysts and methylene blue. Measurements were taken with catalysts suspensions in the same experimental conditions than the reaction tests.

4. Conclusions

Atomic layer deposition was used for adding SiO_2 to the surface of P25. Due to the incorporation of the silica layer, a loss of photocatalytic efficiency took place as observed in the results obtained for the oxidation of methanol. However, the incorporation of SiO_2 also improves the adhesion of bacteria to the surface of the material, partially counteracting the decrease in the photocatalytic activity and leading to a comparable global inactivation rate. Moreover, the SiO_2 layer improves significantly the adsorption of methylene blue, increasing the global removal rate with respect to the bare P25 material as a results of the combination of adsorption and photocatalytic oxidation.

On the other hand, both catalysts show only a small loss of photocatalytic efficiency when immobilized in a macroporous foam as catalytic support. This 3D catalytic system provides the next step towards implementation of the photocatalytic process in a reusable structured reactor configuration without the necessity of a recovery stage of the dispersed catalyst particles.

Author Contributions: Methodology, M.M.-S. and D.B.; data curation, J.M. and J.R.v.O.; writing—original draft preparation, M.M.-S.; writing—review and editing, M.M.-S., D.B., J.R.v.O. and J.M.; supervision, J.M. and J.R.v.O.; funding acquisition, J.M. and J.R.v.O. All authors have read and agreed to the published version of the manuscript.

Funding: This research is supported by the TU Delft | Global Initiative, a program of the Delft University of Technology to boost Science and Technology for Global Development. The authors gratefully acknowledge the financial support of the Spanish State Research Agency (AEI) and the Spanish Ministry of Science and Universities through the CALYPSOL-ATECWATER project (RTI2018-097997-B-C33) and Comunidad de Madrid through the program REMTAVARES (P2018/EMT-4341). Dr. Martín-Sómer also acknowledges MECD for its FPU grant (FPU014/04389) and Universidad Rey Juan Carlos for funding his research stay at the TU Delft.

Conflicts of Interest: The authors declare no conflict of interest.

References

1. Ohno, T.; Sarukawa, K.; Tokieda, K.; Matsumura, M. Morphology of a TiO_2 photocatalyst (Degussa, P25) consisting of anatase and rutile crystalline phases. *J. Catal.* **2001**, *203*, 82–86. [CrossRef]
2. Daghrir, R.; Drogui, P.; Robert, D. Modified TiO_2 for environmental photocatalytic applications: A review. *Ind. Eng. Chem. Res.* **2013**, *52*, 3581–3599. [CrossRef]
3. Marschall, R.; Wang, L. Non-metal doping of transition metal oxides for visible-light photocatalysis. *Catal. Today* **2014**, *225*, 111–135. [CrossRef]
4. Pelaez, M.; Nolan, N.T.; Pillai, S.C.; Seery, M.K.; Falaras, P.; Kontos, A.G.; Dunlop, P.S.M.; Hamilton, J.W.J.; Byrne, J.A.; O'Shea, K.; et al. A review on the visible light active titanium dioxide photocatalysts for environmental applications. *Appl. Catal. B Environ.* **2012**, *125*, 331–349. [CrossRef]
5. Rehman, S.; Ullah, R.; Butt, A.M.; Gohar, N.D. Strategies of making TiO_2 and ZnO visible light active. *J. Hazard. Mater.* **2009**, *170*, 560–569. [CrossRef]
6. Osman, Y.; Jamal, R.; Rahman, A.; Xu, F.; Ali, A.; Abdiryim, T. Comparative study on poly(3,4-propylenedioxythiophene)/TiO_2 nanocomposites synthesized by mechanochemical and chemical solution methods. *Synth. Met.* **2013**, *179*, 54–59. [CrossRef]
7. Cheng, L.; Qiu, S.; Chen, J.; Shao, J.; Cao, S. A practical pathway for the preparation of Fe_2O_3 decorated TiO_2 photocatalyst with enhanced visible-light photoactivity. *Mater. Chem. Phys.* **2017**, *190*, 53–61. [CrossRef]
8. Li, X.; Lin, H.; Chen, X.; Niu, H.; Liu, J.; Zhang, T.; Qu, F. Dendritic α-Fe_2O_3//TiO_2 nanocomposites with improved visible light photocatalytic activity. *Phys. Chem. Chem. Phys.* **2016**, *18*, 9176–9185. [CrossRef]
9. Cheng, G.; Xu, F.; Xiong, J.; Wei, Y.; Stadler, F.J.; Chen, R. A novel protocol to design TiO_2-Fe_2O_3 hybrids with effective charge separation efficiency for improved photocatalysis. *Adv. Powder Technol.* **2017**, *28*, 665–670. [CrossRef]
10. Michalchuk, A.A.L.; Tumanov, I.A.; Konar, S.; Kimber, S.A.J.; Pulham, C.R.; Boldyreva, E.V. Challenges of mechanochemistry: Is in situ real-time quantitative phase analysis always reliable? A case study of organic salt formation. *Adv. Sci.* **2017**, *4*, 1700132. [CrossRef]
11. Ritala, M.; Leskelä, M. Atomic layer deposition. *Handb. Thin Film.* **2002**, *409*, 103–159.
12. Grillo, F.; Kreutzer, M.T.; van Ommen, J.R. Modeling the precursor utilization in atomic layer deposition on nanostructured materials in fluidized bed reactors. *Chem. Eng. J.* **2015**, *268*, 384–398. [CrossRef]

13. van Ommen, J.R.; Goulas, A. Atomic layer deposition on particulate materials. *Mater. Today Chem.* **2019**, *14*, 100183. [CrossRef]
14. Valencia, S.; Marín, J.M.; Restrepo, G. Study of the bandgap of synthesized titanium dioxide nanoparticules using the sol-gel method and a hydrothermal treatment. *Open Mater. Sci. J.* **2010**, *4*, 9–14. [CrossRef]
15. Kalpaklı, Y.K.; Akgun, M.; Köneçoğlu, G.; Toygun, Ş.; Kalpaklı, Y.; Akgün, M. Photocatalytic degradation of textile dye CI Basic Yellow 28 wastewater by Degussa P25 based TiO_2. *Adv. Environ. Res.* **2015**, *4*, 25–38.
16. Martín, A.; Morales, V.; Ortiz-Bustos, J.; Pérez-Garnes, M.; Bautista, L.F.; García-Muñoz, R.A.; Sanz, R. Modelling the adsorption and controlled release of drugs from the pure and amino surface-functionalized mesoporous silica hosts. *Microporous Mesoporous Mater.* **2018**, *262*, 23–34. [CrossRef]
17. Martín-Sómer, M.; Vega, B.; Pablos, C.; van Grieken, R.; Marugán, J. Wavelength dependence of the efficiency of photocatalytic processes for water treatment. *Appl. Catal. B Environ.* **2018**, *221*, 258–265. [CrossRef]
18. Houas, A.; Lachheb, H.; Ksibi, M.; Elaloui, E.; Guillard, C.; Herrmann, J.M. Photocatalytic degradation pathway of methylene blue in water. *Appl. Catal. B Environ.* **2001**, *31*, 145–157. [CrossRef]
19. Soltani, T.; Entezari, M.H. Photolysis and photocatalysis of methylene blue by ferrite bismuth nanoparticles under sunlight irradiation. *J. Mol. Catal. A Chem.* **2013**, *377*, 197–203. [CrossRef]
20. McCaughan, B.; Rouanet, C.; Fowley, C.; Nomikou, N.; McHale, A.P.; McCarron, P.A.; Callan, J.F. Enhanced ROS production and cell death through combined photo- and sono-activation of conventional photosensitising drugs. *Bioorg. Med. Chem. Lett.* **2011**, *21*, 5750–5752. [CrossRef]
21. May, J.M.; Qu, Z.C.; Whitesell, R.R. Generation of oxidant stress in cultured endothelial cells by methylene blue: Protective effects of glucose and ascorbic acid. *Biochem. Pharmacol.* **2003**, *66*, 777–784. [CrossRef]
22. Marugán, J.; van Grieken, R.; Pablos, C.; Sordo, C. Analogies and differences between photocatalytic oxidation of chemicals and photocatalytic inactivation of microorganisms. *Water Res.* **2010**, *44*, 789–796. [CrossRef] [PubMed]
23. Jouali, A.; Salhi, A.; Aguedach, A.; Aarfane, A.; Ghazzaf, H.; Lhadi, E.K.; el krati, M.; Tahiri, S. Photo-catalytic degradation of methylene blue and reactive blue 21 dyes in dynamic mode using TiO_2 particles immobilized on cellulosic fibers. *J. Photochem. Photobiol. A Chem.* **2019**, *383*, 112013. [CrossRef]
24. Pablos, C.; Marugán, J.; van Grieken, R.; Adán, C.; Riquelme, A.; Palma, J. Correlation between photoelectrochemical behaviour and photoelectrocatalytic activity and scaling-up of P25-TiO_2 electrodes. *Electrochim. Acta* **2014**, *130*, 261–270. [CrossRef]
25. Marugán, J.; van Grieken, R.; Sordo, C.; Cruz, C. Kinetics of the photocatalytic disinfection of *Escherichia coli* suspensions. *Appl. Catal. B Environ.* **2008**, *82*, 27–36. [CrossRef]
26. Xiong, J.; Li, G.; Hu, C. Treatment of methylene blue by mesoporous Fe/SiO_2 prepared from rice husk pyrolytic residues. *Catal. Today.* **2019**. [CrossRef]
27. Nishikawa, H.; Takahara, Y. Adsorption and photocatalytic decomposition of odor compounds containing sulfur using TiO_2/SiO_2 bead. *J. Mol. Catal. A Chem.* **2001**, *172*, 247–251. [CrossRef]
28. Yaparatne, S.; Tripp, C.P.A. Amirbahman, Photodegradation of taste and odor compounds in water in the presence of immobilized TiO_2-SiO_2 photocatalysts. *J. Hazard. Mater.* **2018**, *346*, 208–217. [CrossRef]
29. Uddin, M.T.; Islam, M.A.; Mahmud, S.; Rukanuzzaman, M. Adsorptive removal of methylene blue by tea waste. *J. Hazard. Mater.* **2009**, *164*, 53–60. [CrossRef]
30. Li, J.; Zhou, S.L.; Hong, G.B.; Chang, C.T. Hydrothermal preparation of P25-graphene composite with enhanced adsorption and photocatalytic degradation of dyes. *Chem. Eng. J.* **2013**, *219*, 486–491. [CrossRef]
31. Nguyen, C.H.; Juang, R.S. Efficient removal of methylene blue dye by a hybrid adsorption–photocatalysis process using reduced graphene oxide/titanate nanotube composites for water reuse. *J. Ind. Eng. Chem.* **2019**, *76*, 296–309. [CrossRef]
32. Ngoh, Y.S.; Nawi, M.A. Fabrication and properties of an immobilized P25 TiO_2-montmorillonite bilayer system for the synergistic photocatalytic-adsorption removal of methylene blue. *Mater. Res. Bull.* **2016**, *76*, 8–21. [CrossRef]
33. Jucker, B.A.; Harms, H.; Hug, S.J.; Zehnder, A.J.B. Adsorption of bacterial surface polysaccharides on mineral oxides is mediated by hydrogen bonds. *Colloids Surf. B Biointerfaces* **1997**, *9*, 331–343. [CrossRef]
34. Huang, T.T.; Sturgis, J.; Gomez, R.; Geng, T.; Bashir, R.; Bhunia, A.K.; Robinson, J.P.; Ladisch, M.R. Composite surface for blocking bacterial adsorption on protein biochips. *Biotechnol. Bioeng.* **2003**, *81*, 618–624. [CrossRef] [PubMed]

35. Erdural, B.; Bolukbasi, U.; Karakas, G. Photocatalytic antibacterial activity of TiO_2-SiO_2 thin films: The effect of composition on cell adhesion and antibacterial activity. *J. Photochem. Photobiol. A Chem.* **2014**, *283*, 29–37. [CrossRef]
36. Martín-Sómer, M.; Pablos, C.; de Diego, A.; van Grieken, R.; Encinas, Á.; Monsalvo, V.M.; Marugán, J. Novel macroporous 3D photocatalytic foams for simultaneous wastewater disinfection and removal of contaminants of emerging concern. *Chem. Eng. J.* **2019**, *366*, 449–459. [CrossRef]
37. Grillo, F.; Moulijn, J.A.; Kreutzer, M.T.; van Ommen, J.R. Nanoparticle sintering in atomic layer deposition of supported catalysts: Kinetic modeling of the size distribution. *Catal. Today* **2018**, *316*, 51–61. [CrossRef]
38. Guo, J.; van Bui, H.; Valdesueiro, D.; Yuan, S.; Liang, B.; van Ommen, J. Suppressing the photocatalytic activity of TiO_2 nanoparticles by extremely thin Al_2O_3 films grown by gas-phase deposition at ambient conditions. *Nanomaterials* **2018**, *8*, 61. [CrossRef]
39. Philippe, K.K.; Timmers, R.; van Grieken, R.; Marugan, J. Photocatalytic disinfection and removal of emerging pollutants from effluents of biological wastewater treatments, using a newly developed large-scale solar simulator. *Ind. Eng. Chem. Res.* **2016**, *55*, 2952–2958. [CrossRef]

Article

Electrochemical Oxidation of an Organic Dye Adsorbed on Tin Oxide and Antimony Doped Tin Oxide Graphene Composites

Farbod Sharif and Edward P. L. Roberts *

University of Calgary, Department of Chemical and Petroleum Engineering, 2500 University Drive NW, Calgary, AB T2N 1N4, Canada; shariff@ucalgary.ca

* Correspondence: edward.roberts@ucalgary.ca; Tel.: +1-403-220-4466

Received: 4 February 2020; Accepted: 18 February 2020; Published: 21 February 2020

Abstract: Electrochemical regeneration suffers from low regeneration efficiency due to side reactions like oxygen evolution, as well as oxidation of the adsorbent. In this study, electrically conducting nanocomposites, including graphene/SnO_2 (G/SnO_2) and graphene/Sb-SnO_2 (G/Sb-SnO_2) were successfully synthesized and characterized using nitrogen adsorption, scanning electron microscopy, transmission electron microscopy, and Raman spectroscopy. Thereafter, the adsorption and electrochemical regeneration performance of the nanocomposites were tested using methylene blue as a model contaminant. Compared to bare graphene, the adsorption capacity of the new composites was ≥40% higher, with similar isotherm behavior. The adsorption capacity of G/SnO_2 and G/Sb-SnO_2 were effectively regenerated in both NaCl and Na_2SO_4 electrolytes, requiring as little charge as 21 C mg^{-1} of adsorbate for complete regeneration, compared to >35 C mg^{-1} for bare graphene. Consecutive loading and anodic electrochemical regeneration cycles of the nanocomposites were carried out in both NaCl and Na_2SO_4 electrolytes without loss of the nanocomposite, attaining high regeneration efficiencies (ca. 100%).

Keywords: graphene; tin oxide; antimony doped tin oxide; adsorption; electrochemical oxidation

---

1. Introduction

Adsorption is a promising method for the removal of soluble and insoluble organics from wastewater effluents due to ease of operation, a wide range of applications, and a high level of purity of the treated water [1]. However, handling of the loaded adsorbent is a challenge because of factors that constrain disposal, including the toxicity of the adsorbate and the high cost of replacement adsorbent [2,3]. Electrochemical regeneration has shown potential for effectively recovering the adsorptive capacity of the loaded adsorbents [4–7].

Early studies of electrochemical regeneration with an activated carbon adsorbent showed that although it has a high adsorptive capacity, it requires long regeneration times, resulting in high energy consumption. This is mainly due to the porous surface and low conductivity of the activated carbon [8–10]. Brown et al. [11–13] studied an alternative material, graphite intercalation compound (GIC), which is nonporous and has high electrical conductivity. This material demonstrated high regeneration efficiency but low adsorptive capacity. In the previous study [14], reduced graphene oxide (rGO)/magnetite was chosen as an adsorbent since it has a nonporous surface, high surface area, and high electrical conductivity. The findings revealed that rGO/magnetite has satisfactory adsorptive capacity which can be completely regenerated. However, the electrochemical regeneration process caused oxidation and corrosion of the adsorbent. Similar adsorbent oxidation was also seen by Nkrumah-Amoako et al. [15] for a GIC adsorbent. Graphitic materials display behaviors of both active and inactive electrodes, but the dominant mechanism for organic oxidation is believed to be direct electron transfer [16]. This is likely the main reason for oxidation of the graphene as well.

It is essential to tackle the problem of adsorbent oxidation, and this can be done by changing the dominant mechanism from direct electron transfer to mediated degradation using electro-generated hydroxyl radicals. One approach would be to coat the surface of the carbon adsorbent, i.e., graphene, with materials that behave as inactive anodes, such as boron-doped diamond (BDD), SnO_2, Sb-SnO_2, or PbO_2. Due to the high overpotential of the inactive materials for oxygen evolution, they are considered good candidates for organic oxidation by means of reactive oxygen species such as hydroxyl radicals. Extensive research has been conducted on inactive materials to increase the rate of reactive oxygen species production. Ozone production on Sb-SnO_2/Ti, hydroxyl radical generation on fluoride-doped lead oxide, hydrogen peroxide generation on niobium lead oxide, and hydroxyl radical generation on BDD and TiO_2/Ti are several good examples [17–23]. These materials increase the oxygen evolution overpotential, minimizing this side reaction, thus increasing the current efficiency for organic oxidation [21,24,25]. Owing to its large band gap of 5.45 eV, BDD is the best inactive, corrosion-resistant material [26]; however, due to its high cost, BDD cannot be effectively used for adsorption and electrochemical regeneration. PbO_2 also has a high overpotential for oxygen evolution [27] as well as a much lower cost than BDD, but leaching of lead from the PbO_2 into the solution can contaminate the treated water [28].

In the current study, graphene/SnO_2 (G/SnO_2) and graphene/Sb-SnO_2 (G/Sb-SnO_2) nanocomposites were prepared, characterized, and utilized in an adsorption and electrochemical regeneration process. Our previous study on electrochemical regeneration of graphene TiO_2 adsorbents materials revealed that the addition of the TiO_2 nanoparticles to the surface of the graphene increases catalytic activity, reducing the required time for complete regeneration, and also protects the graphene surface from corrosion [29]. However, the prepared adsorbents did not demonstrate good adsorptive capacity due to loss of amorphous carbon during the calcination, required for synthesis of the TiO_2 nanocomposite.

The goal of the present study was to prepare new materials with higher adsorptive capacity and comparable catalytic activity to the graphene/TiO_2 nanocomposite which can be applied in successive adsorption electrochemical regeneration process. In addition to the methylene blue (MB) adsorption and electrochemical regeneration behavior of these adsorbents, their electrochemical properties were also investigated. The effect of the nanomaterial loading on the nanocomposites and the electrolyte types on the electrochemical regeneration and the durability of the new materials was also assessed.

2. Result and Discussion

2.1. Raman Spectra

The surface chemical composition of the graphene and composite samples was characterized by Raman spectroscopy, as shown in Figure 1. The peaks observed at 1358 and 1580 cm^{-1} can be attributed to the D band and G band of the graphene, respectively [30]. The peak at a Raman shift of 572 cm^{-1} can be assigned to the cassiterite SnO_2 nanoparticles; this peak is only present for nanometer-scale SnO_2 particles [31,32]. The four broad Raman peaks at 458, 566, 720, and 632 cm^{-1} can be attributed to SnO_2 nanoparticles doped with antimony. The first three peaks are the surface vibration modes [33], and the last one is the vibration in the plane perpendicular to the c-axis [34]. Thus, the Raman spectra confirm the presence of SnO_2 and Sb-SnO_2 on the nanocomposite materials.

2.2. SEM and TEM

The morphology of the G/SnO_2 and G/Sb-SnO_2 nanocomposites were observed using TEM and SEM (Figures 2 and 3). A comparison between the SEM image of bare graphene, G/SnO_2, and G/Sb-SnO_2 reveals that SnO_2 nanoparticles have adhered to the surface of the graphene; however, due to the SEM limitations, it is not possible to make more quantitative comments on the nanoparticle distribution. The TEM images, Figure 3b,d show that the average sizes of SnO_2 and Sb-SnO_2 nanoparticles are around

5 and 30 nm, respectively. It can also be inferred that the nanoparticles have not been distributed uniformly on the surface; they tend to aggregate due to their high surface energy [35].

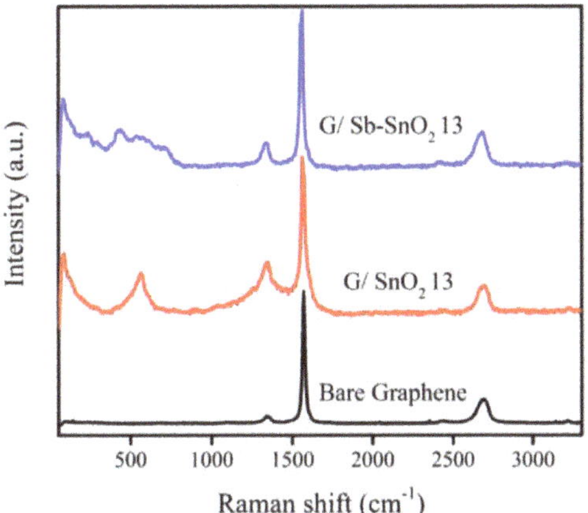

Figure 1. Raman spectra of Bare graphene, SnO_2, Sb-SnO_2, G/SnO_2, and G/Sb-SnO_2 nanocomposites.

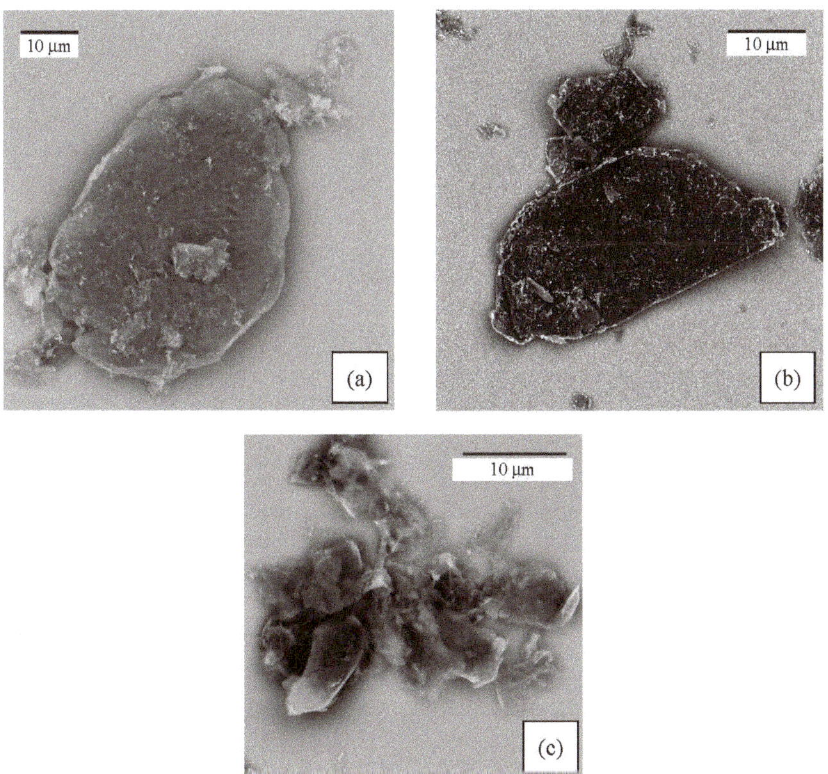

Figure 2. SEM images of the (**a**) G/SnO_2 nanocomposite, (**b**) G/Sb-SnO_2 nanocomposite, and (**c**) bare graphene.

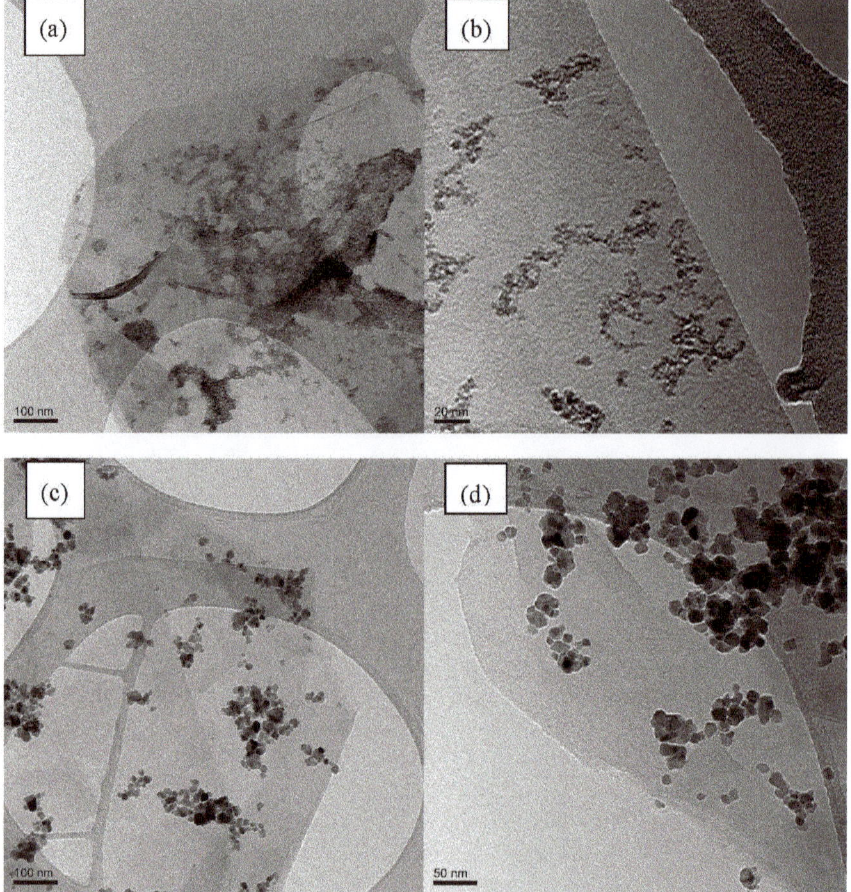

Figure 3. (**a**) Low magnification TEM image of the G/SnO2 nanocomposite (**b**) High-magnification TEM image of the G/SnO2 nanocomposite. (**c**) Low-magnification TEM image of the G/Sb-SnO2 nanocomposite. (**d**) High-magnification TEM image of the G/Sb-SnO2 nanocomposite.

2.3. Adsorption Study

Adsorption isotherm studies were carried out to find the adsorptive capacity of the prepared samples and to determine the adsorption range to be used for adsorbent regeneration studies. The results obtained are depicted in Figure 4. Table 1 shows the calculated surface area for each sample using the BET method. It can be seen that G/SnO_2 and G/Sb-SnO_2 have higher adsorptive capacities than the bare graphene. As the surface area of the graphene and composite materials are all similar, we can conclude that the surface area is not responsible for higher uptake capacity of the composite adsorbents. The difference in the adsorptive capacity can be either due to contribution of the metal oxide in adsorption or the better dispersion of the nanocomposite in the water (i.e., agglomeration of the bare graphene particles may lead to a loss of adsorptive capacity).

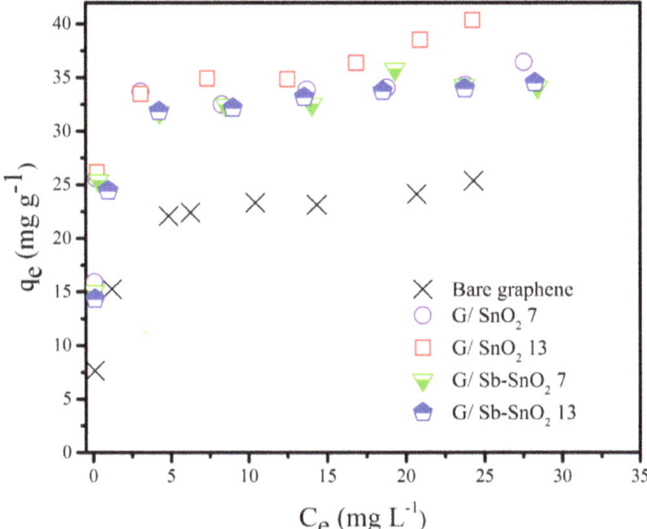

Figure 4. Equilibrium adsorption of methylene blue (MB) on bare graphene, G/SnO_2 7, G/SnO_2 13, G/Sb-SnO_2 7, and G/Sb-SnO_2 13.

Table 1. Specific surface area of the graphene loaded with metal oxide.

Sample	Bare Graphene	G/SnO_2 7	G/SnO_2 13	G/Sb-SnO_2 7	G/Sb-SnO_2 13
Surface area (BET) $m^2\ g^{-1}$	70	84	89	72	69
Surface area (MB) $m^2\ g^{-1}$	62	88	98	85	85

The Langmuir and Freundlich adsorption isotherm models were fitted to the experimental data for MB adsorption on bare graphene, G/SnO_2 7, G/SnO_2 13, G/Sb-SnO_2 7, and G/Sb-SnO_2 13. The Langmuir and Freundlich isotherms were used to evaluate the equilibrium conditions, and they can be expressed by the mathematical Equations (1) and (2), respectively.

$$\text{Langmuir isotherm} \qquad q_e = \frac{K_l b C_e}{1 + bC_e} \tag{1}$$

$$\text{Freundlich isotherm:} \qquad q_e = K_f {C_e}^n \tag{2}$$

where q_e is the loading of adsorbate on the adsorbent in equilibrium with a solution concentration of C_e, K_L and b are the Langmuir isotherm constants, and K_F and n are the Freundlich isotherm constants.

Tables 2–4 show the isotherm constants and the determination coefficient (R^2) obtained by nonlinear regression for each adsorbent. By comparing the value of R^2 for each adsorbent, it can be observed that Langmuir isotherm provides a better fit to the experimental data than the Freundlich isotherm model. Separation factor (R_L) [36] (Equation (3)) and the magnitude of n [37] are important parameters in the Langmuir and Freundlich isotherm as they can indicate whether the adsorption is favorable or not. Area value of R_L or n of less than one indicates favorable adsorption.

$$R_L = \frac{1}{1 + bC_0} \tag{3}$$

Table 2. Langmuir constants for adsorption of MB on Bare graphene, G/SnO_2 7, G/SnO_2 13, G/Sb-SnO_2 7, and G/Sb-SnO_2 13.

	Langmuir Parameters: $q_e = \frac{KbC_e}{1+bC_e}$				
Sample	**Maximum Adsorption ($mg\ g^{-1}$)**	**K**	**b**	**R^2**	$R_L = \frac{1}{1+bC_0}$
Bare graphene	25	24.3	2.28	0.92	0.017
G/SnO_2 7	36	34.53	11.92	0.97	0.003
G/SnO_2 13	40	37.02	8.16	0.94	0.005
G/Sb-SnO_2 7	35	33.82	8.87	0.97	0.005
G/Sb-SnO_2 13	35	33.47	5.19	0.94	0.008

Table 3. Dimensionless Langmuir constants (R_L) for adsorption of MB on bare graphene at low and high concentrations for different adsorbents.

$R_L = \frac{1}{1+bC_0}$ at Two Different Initial Concentrations		
	10	50
Bare graphene	0.04	0.009
G/ SnO_2 7	0.008	0.002
G/ SnO_2 13	0.012	0.002
G/ Sb-SnO_2 7	0.011	0.002
G/ Sb-SnO_2 13	0.019	0.004

Table 4. Freundlich constants for adsorption of MB on bare graphene, G/SnO_2 7, G/SnO_2 13, G/Sb-SnO_2 7, and G/Sb-SnO_2 13.

	Freundlich Parameters: $q_e = KC_e^n$			
	Maximum Adsorption ($mg\ g^{-1}$)	**K**	**n**	**R^2**
Bare graphene	25	14.90	0.17	0.92
G/SnO_2 7	36	26.44	0.09	0.83
G/ SnO_2 13	40	26.63	0.12	0.89
G/Sb-SnO_2 7	35	25.01	0.11	0.88
G/ Sb-SnO_2 13	35	23.39	0.13	0.91

It can be seen from the R_L values in Table 3 that the adsorption process of MB on graphene nanocomposites is favorable at all concentrations, particularly at low concentrations. The values of n obtained for the Freundlich isotherm are consistent with the with R_L values.

The surface area of the nanocomposite can be estimated using MB. As the adsorption of MB on the nanocomposites follows the Langmuir isotherm, it can be concluded that a monolayer of MB has been adsorbed on the surface. It has been reported that one molecule of MB can occupy 130 Å^2 on the surface of the adsorbent [38,39]. Table 1 shows the surface area for each adsorbent. It can be seen surface area calculated using BET and MB are more or less the same.

2.4. Electrochemical Regeneration

Figure 5a,b and Figure S1a,b demonstrate the evolution of the regeneration efficiency with the charge passed through the electrolytic cell using NaCl and Na_2SO_4 electrolytes. For all of the adsorbents in both electrolytes, the regeneration efficiency shows a steep increase as the charge was increased from 0 to 1000 C g^{-1}, and then it slowly increases until it reaches complete regeneration. The findings reveal that the adsorptive capacity of all of the adsorbents can be completely restored.

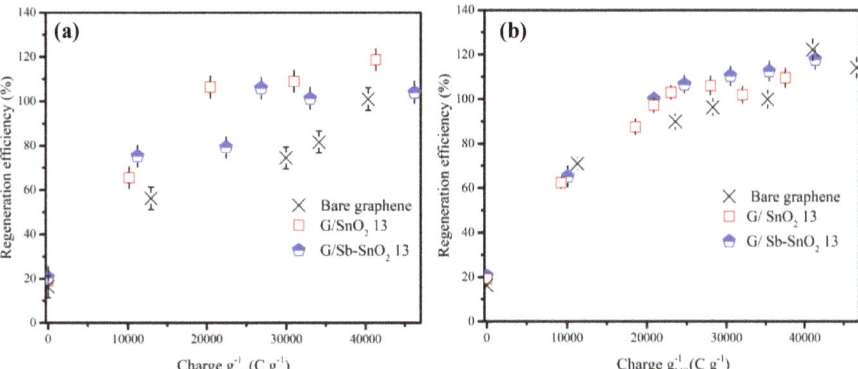

Figure 5. Effect of charge passed on regeneration efficiency of MB adsorption on bare graphene, G/SnO_2 13, and $G/Sb\text{-}SnO_2$ 13. (**a**) NaCl electrolyte and (**b**) Na_2SO_4 electrolyte (current density of 10 mA cm^{-2}).

In previous studies, it has been reported that the utilization of NaCl instead of Na_2SO_4 as the electrolyte in the regeneration cell leads to higher regeneration efficiency [8,40]. However, the results of this study suggest that the choice of electrolyte has minimal impact on the regeneration efficiency. Moreover, with sodium sulfate as the electrolyte, if more charge is passed through the cell after complete regeneration of the spent adsorbents, the adsorptive capacity increases less compared to regeneration with a sodium chloride electrolyte. Previous studies have indicated that electrochemical regeneration efficiencies of greater than 100% are associated with oxidation of graphite adsorbent surfaces [12,15]. Thus, the previous studies suggest that with sodium sulfate, less oxidation of the adsorbent's surface occurs, allowing it to remain almost intact. A secondary benefit of using the sodium sulfate electrolyte is that it can hinder the formation of toxic chlorinated hydrocarbons [11].

Previous work has shown that complete regeneration of graphene TiO_2 nanocomposite loaded with MB dye could be achieved by passing a charge of 21 C mg^{-1} through the cell [29]. The two main problems associated with this nanocomposite compared to bare graphene were lower adsorptive capacity and a higher cell voltage required for electrochemical regeneration. As explained previously, G/SnO_2 and $G/Sb\text{-}SnO_2$ showed higher adsorptive capacity compared to bare graphene.

Table 5 shows the specific charge required for complete regeneration of the loaded adsorbents. The prepared nanocomposites had a required charge for complete oxidation as low as 21 C mg^{-1} for G/SnO_2 using NaCl as an electrolyte and for $G/Sb\text{-}SnO_2$ using Na_2SO_4 as an electrolyte, which is comparable to the graphene TiO_2 nanocomposite results previously reported. Energy consumption for 100% regeneration of the loaded adsorbents is calculated by multiplying the required charge by the cell voltage. As shown in Table 5, the G/SnO_2 and $G/Sb\text{-}SnO_2$ nanocomposites demonstrate lower charge requirements for complete regeneration than bare graphene using NaCl or Na_2SO_4 as the electrolyte, with similar cell voltages (≈2.6 V). Thus the G/SnO_2 and $G/Sb\text{-}SnO_2$ adsorbents require less energy for complete regeneration. These results confirm the higher electrocatalytic activity of the G/SnO_2 and $G/Sb\text{-}SnO_2$ nanocomposites for regeneration compared to bare graphene.

Table 5. Specific charge required for 100% regeneration efficiency of bare graphene, G/SnO_2 7, G/SnO_2 13, $G/Sb\text{-}SnO_2$ 7, and $G/Sb\text{-}SnO_2$ 13 adsorbents loaded with MB, (regeneration at 10 mA cm^{-2}).

Adsorbent:	Bare Graphene	G/ SnO_2 7	G/ SnO_2 13	G/Sb-SnO_2 7	G/Sb-SnO_2 13
Charge passed NaCl (C mg^{-1})	39	21	21	27	27
Charge passed Na_2SO_4 (C mg^{-1})	35	23	23	21	21
Cell Voltage (V)	≈2.6	≈2.6	≈2.6	≈2.6	≈2.6

2.5. Adsorption/Regeneration Cycles

Multiple loading and electrochemical regeneration cycles were carried out to investigate the reusability of the G/SnO_2 and G/Sb-SnO_2 nanocomposites. The adsorbents were utilized to adsorb MB and regenerated for five successive cycles using either 2 wt. % NaCl or Na_2SO_4 as electrolytes and 10 mA cm^{-2} current density. For each adsorbent, the required electrochemical treatment time for 100% regeneration was estimated based on the data plotted in Figure 5a,b and Figure S1a,b Table 6 shows the mass of adsorbent recovered after each cycle for the five different adsorbent materials. For the G/SnO_2 and G/Sb-SnO_2 nanocomposites, only around 10% of the mass of adsorbent was lost over five cycles of adsorption and regeneration. The loss was within 1% to 2% of the loss observed from a control experiment to determine the physical losses associated with the recovery of the adsorbent after filtration. In contrast, with the bare graphene adsorbent, around 30% of the mass was lost over the five cycles of adsorption and regeneration, confirming that significant corrosive oxidation of the graphene was occurring during regeneration.

Table 6. Mass of the adsorbent recovered after five consecutive cycles of adsorption and regeneration.

Sample	G/SnO_2 7	G/SnO_2 13	G/Sb-SnO_2 7	G/Sb-SnO_2 13	Bare Graphene
Mass	0.91	0.88	0.89	0.90	0.70

As shown in Figure 6a,b, the electrochemical oxidation resulted in no reduction of regeneration efficiency, even after five cycles. Figure 6a,b also shows that the change in regeneration efficiency of the bare graphene was much more than the G/SnO_2 and G/Sb-SnO_2 nanocomposites. This increase in the regeneration efficiency was due to the oxidation of the bare graphene during electrochemical regeneration, which led to the creation of more adsorption sites, either by increasing the surface area or the number of functional groups [15,41]. The adsorptive capacity and consequently regeneration efficiency of the nanocomposites changed less than for bare graphene, which suggests that the addition of the SnO_2 and Sb-SnO_2 provides some protection of the surface of the graphene from oxidation. By comparing Figure 6a,b, it is evident that the regeneration efficiency of the adsorbents in sodium sulfate increased less than that observed with sodium chloride. This result is consistent with the charge loading results (Figure 5a,b, and Figure S1a,b and confirms that there is a greater tendency for surface oxidation of the graphene using NaCl electrolyte.

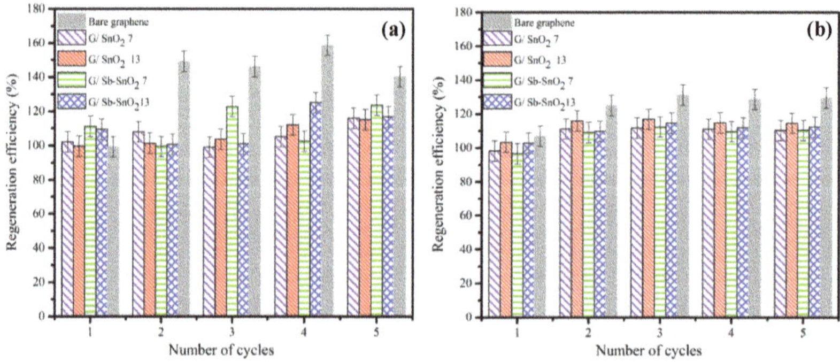

Figure 6. Regeneration efficiency for a series of adsorption and electrochemical regeneration cycles of MB adsorption on bare graphene, G/SnO_2 7, G/SnO_2 13, G/Sb-SnO_2 7, and G/Sb-SnO_2 13 nanocomposites in (**a**) NaCl electrolyte and (**b**) Na_2SO_4 electrolyte. Adsorption was carried out under conditions that give close to the maximum loading of MB on the adsorbent. Regeneration of the bare graphene, G/SnO_2 7, G/SnO_2 13, G/Sb-SnO_2 7, and G/Sb-SnO_2 13 adsorbents were carried out for 14, 10, 10, 12, and 12 min, respectively, at a current density of 10 mA cm^{-2}.

A common major problem associated with using NaCl in an electrochemical water treatment process is the formation of chlorinated compounds, which may be more toxic than the primary pollutants [42]. Thus, in addition to reduced oxidation of the graphene, sodium sulfate can be a lower-risk alternative to sodium chloride for regeneration of G/SnO_2 and G/Sb-SnO_2 nanocomposite adsorbents.

2.6. Electrochemical Characterization of the Adsorbents

2.6.1. Linear Sweep Voltammetry

In order to assess the electrochemical characteristics of the nanocomposites, linear sweep voltammetry (LSV) was carried out. Figure 7a shows the results for bare graphene, G/SnO_2, and G/Sb-SnO_2 nanocomposites, in which the current flow was measured as the applied potential was increased. The experiments were performed in 0.5 M Na_2SO_4 at a scan rate of 100 mVs^{-1}, with Ag/AgCl and platinum wire used as reference and counter electrodes respectively. Onset potentials of 1.65, 1.87, 1.87, 1.9, and 1.87 V were measured for bare graphene, G/SnO_2 7, G/SnO_2 13, G/Sb-SnO_2 7, and G/Sb-SnO_2 13 adsorbents, respectively, for the oxygen evolution reaction. The delay in the onset potential indicates the suppression of this side reaction and increasing the charge efficiency for generating reactive oxygen species. Such a trend has also been observed by different researchers for SnO_2 and Sb-SnO_2 electrocatalyst materials [21,25,43,44].

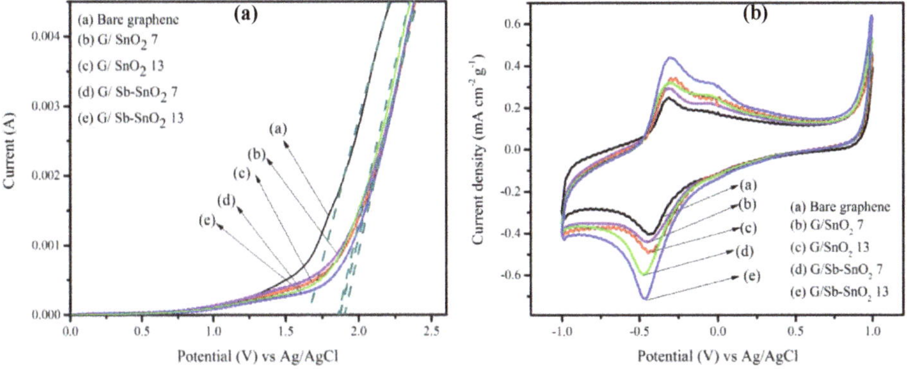

Figure 7. (**a**) Linear sweep voltammograms of (a) bare graphene, (b) G/SnO_2 7, (c) G/SnO_2 3, (d) G/Sb-SnO_2 7, and (e) G/Sb-SnO_2 13. Experiments were conducted using 0.5 mol L^{-1} sodium sulfate at a scan rate of 100 mV S^{-1} (**b**) Cyclic voltammetry of (a) bare graphene, (b) G/SnO_2 7, (c) G/SnO_2 13, (d) G/Sb-SnO_2 7, and (e) G/Sb-SnO_2 13 at a scan rate of 10 mV S^{-1} using a solution containing 25 ppm MB and 0.5 mol L^{-1} sodium sulphate.

2.6.2. Cyclic Voltammetry

Figure 7b shows the cyclic voltammogram of the bare graphene, G/SnO_2, and G/Sb-SnO_2 nanocomposites in 0.5 mol L^{-1} Na_2SO_4 solution containing 25 mg L^{-1} of MB. The current density was normalized by the graphene loading for all of the adsorbents. Unlike the bare graphene, G/SnO_2 and G/Sb-SnO_2 demonstrate a well-defined pair of redox peaks for MB oxidation and reduction. In addition, the peak separation, ΔE_p, was smaller for G/SnO_2 and G/Sb-SnO_2 nanocomposites relative to bare graphene, demonstrating that the electron transfer was faster for these nanocomposites. It can be seen that the SnO_2 and Sb-SnO_2 nanoparticles-coated graphene show a higher oxidation peak than that of bare graphene. This implies that the addition of SnO_2 and Sb-SnO_2 nanoparticles to the graphene increases the electrocatalytic activity, apparently due to synergic effects of the presence of both an n-type semiconductor with a large band gap and graphene [45,46]. It can also be seen that by increasing the nanoparticles loading from 7 to 13 wt. %, the current density increased for both G/SnO_2 and G/Sb-SnO_2 nanocomposites. Furthermore, at the same loading of the nanoparticles, G/Sb-SnO_2 shows better catalytic activity compared to G/SnO_2 nanocomposites, as the Sb-SnO_2 acts similar to a metal electrocatalyst and exhibits a high overpotential of oxygen evolution, which can increase its electrocatalytic activity [47–49].

3. Experimental

Nanocomposite preparation: a known volume of as-received sol of SnO_2 (15 wt. %) or Sb-SnO_2 (20 wt. %) (Nyacol Inc., Ashland, MA, USA) was added to 150 mL of a suspension of graphene (GNPs 25 M, XG sciences Inc., Lansing, MI, USA) in DI water (containing 1 g L^{-1} of graphene) and mixed for 24 h. The mixture was dried at 70 °C for 12 h. In this paper, the nanocomposites are named with respect to their composition: thus G/SnO_2 7 corresponds to a graphene/SnO_2 composite with a loading of 7 wt. % SnO_2. Similarly, G/SnO_2 13 has a loading of 13 wt. % SnO_2, and G/Sb-SnO_2 7 and G/Sb-SnO_2 13 have loadings of 7 wt. % and 13 wt. % Sb-SnO_2, respectively.

Characterization: The morphologies of the prepared nanocomposites were characterized using scanning electron microscopy (SEM) on a Zeiss Supra55 field-emission SEM (Carl Zeiss Microscopy LLC, White Plains, NY, USA) and transmission electron microscopy (TEM) using a Tecnai TF20 G2 FEG-TEM (FEI, Hillsboro, OR, USA) with a 200−kV acceleration voltage. A Witec alpha 300 R Confocal Raman Microscope (Witec GmbH, Ulm, Germany) was utilized to obtain Raman spectra using a 532 nm laser. The surface area was measured with N_2 physisorption (TriStar 3000, Micromeritics Instrument Corp., Norcross, GA, USA) at −196 °C. Before measurement, all samples were degassed at 150 °C for 12 h. The specific surface area was calculated using the Brunauer¬–Emmett–Teller (BET) method in the relative pressure (P/P_o) range of 0.01 e 0.99. The concentration of MB in the synthetic wastewater was measured using UV-Visible absorption spectroscopy (UV-2600, Shimadzu, Columbia, MD, USA) at a wavelength of 664 nm [50].

Adsorption study: to obtain the adsorption isotherm, experiments were carried out via a 'bottle point' method. Briefly, 0.1 g of adsorbent was mixed for 30 minutes with 150 mL of the MB solution with different concentrations at room temperature. Synthetic wastewater was prepared using deionized water and MB only. The treated water was filtered, and the filtrate concentration was measured using UV-Visible spectrophotometer.

Electrochemical regeneration: an electrolytic cell was used to regenerate the spent adsorbents using a graphite plate as the anode and a stainless-steel plate as the cathode, as described in an earlier study [14]. Sodium chloride and sodium sulfate (VWR, Radnor, PA, USA) were used as electrolytes. A constant current density of 10 mA cm^{-2} was applied to the cell using a Metrohm Autolab PGSTAT potentiostat (Metrohm AG, Herisaum, Switzerland). The regeneration efficiency was calculated as the ratio of the adsorptive capacity after regeneration to the initial adsorptive capacity, each measured under the same adsorption conditions [51,52].

Electrochemical properties: electrochemical measurements, including cyclic voltammetry (CV) and linear sweep voltammetry (LSV) were performed using the Autolab PGSTAT potentiostat. A volume of 100 mL of a 0.5 mol L^{-1} Na_2SO_4 containing 25 ppm MB (VWR, Radnor, PA, USA) was used as an electrolyte. Modified glassy carbon with nanocomposites, Ag/AgCl, and platinum wire were used as working, reference, and counter electrodes, respectively. The modified glassy carbon was prepared by drop-casting a suspension of the nanocomposites (1 mg of adsorbent in 1 mL of Nafion solution, from IonPower, New Castle, DE, USA, with a Nafion-to-adsorbent mass ratio of ~0.1) on a glassy carbon electrode and drying at 70 °C. CV was carried out in a potential range of −1.0 V to +1.0 V at a scanning rate of 10 mV S^{-1}.

4. Conclusions

This study reported the application of G/SnO_2 and G/Sb-SnO_2 nanocomposites in an adsorption and electrochemical regeneration process, using both the high conductivity of the graphene and the large band gap of the tin oxide to improve regeneration efficiency. The adsorptive capacity obtained for the nanocomposites was ≥35 mg g^{-1}, which is approximately 1.4- and 1.75-fold higher than the bare graphene and previously reported graphene titanium oxide [29], respectively.

All the prepared nanocomposites showed the ability to attain 100% regeneration efficiency in both NaCl and Na_2SO_4 electrolytes. With the assistance of SnO_2 and Sb-SnO_2, the charge efficiency of the process was significantly improved. The results indicate that the required charge passed for complete oxidation of the MB decreased by 50% and 30% for G/SnO_2 and G/Sb-SnO_2 in an NaCl electrolyte and 35% and 40% for G/SnO_2 and G/Sb-SnO_2 in an Na_2SO_4 electrolyte. The reason could be the higher oxygen evolution onset potential along with the higher activity of the SnO_2 and Sb-SnO_2. Although the cyclic voltammetry results suggested that G/Sb-SnO_2 has higher catalytic activity than G/SnO_2, the regeneration results show similar charge requirements for complete regeneration for both G/SnO_2 and G/Sb-SnO_2 nanocomposites. The higher catalytic activity of the Sb-SnO_2-modified graphene observed is consistent with previous studies that have demonstrated the suitability of this for electrocatalysis for advanced oxidation combined with a high oxygen overpotential [46,47,50]. An advantage of G/SnO_2 and G/Sb-SnO_2 over TiO_2/graphene nanocomposites is that they have less of an impact on the electrical conductivity of the graphene, minimizing the ohmic losses during regeneration, reducing the energy required.

In addition to the nanocomposites having better performance if Na_2SO_4 is used as the regeneration electrolyte, using a chloride-free electrolyte reduces the risk of creating toxic breakdown products. This makes sodium sulfate a good electrolyte substitute for sodium chloride.

The possibility of formation of breakdown products that may be released into the treated water was not investigated in this study, and further work is needed to explore this aspect. In addition, the effect of the multiple cycles of regeneration on the structure and surface properties of the nanocomposite adsorbent and the lifetime of the adsorbent should also be investigated.

Supplementary Materials: The following are available online at http://www.mdpi.com/2073-4344/10/2/263/s1, Figure S1. Effect of charge passed on regeneration efficiency of MB adsorption on bare graphene, G/SnO_2 7 and G/Sb-SnO_2 7, and (a) NaCl electrolyte and (b) Na_2SO_4 electrolyte (current density of 10 mA cm^2).

Author Contributions: E.P.L.R.: conceptualization, supervision, funding acquisition, project administration, writing—review and editing; F.S.: formal analysis, investigation, methodology, writing original draft, writing—review and editing. All authors have read and agreed to the published version of the manuscript.

Funding: This research has received financial support from the Natural Sciences and Engineering Research Council of Canada (NSERC 435634-2013 and RGPIN-2018-03725) and the Canada Foundation for Innovation (CFI 32613).

Conflicts of Interest: The authors declare no conflict of interest. The funders had no role in the design of the study; in the collection, analyses, or interpretation of data; in the writing of the manuscript, or in the decision to publish the results.

References

1. Ali, I. Water treatment by adsorption columns: Evaluation at ground level. *Sep. Purif. Rev.* **2014**, *43*, 175–205. [CrossRef]
2. Lin, S.H.; Cheng, M.J. Adsorption of phenol and m-chlorophenol on organobentonites and repeated thermal regeneration. *Waste Manag.* **2002**, *22*, 595–603. [CrossRef]
3. Namasivayam, C.; Sangeetha, D. Recycling of agricultural solid waste, coir pith: Removal of anions, heavy metals, organics and dyes from water by adsorption onto $ZnCl_2$ activated coir pith carbon. *J. Hazard. Mater.* **2006**, *135*, 449–452. [CrossRef] [PubMed]
4. Berenguer, R.; Marco-Lozar, J.P.; Quijada, C.; Cazorla-Amorós, D.; Morallón, E. Electrochemical regeneration and porosity recovery of phenol-saturated granular activated carbon in an alkaline medium. *Carbon N. Y.* **2010**, *48*, 2734–2745. [CrossRef]
5. Wang, L.; Balasubramanian, N. Electrochemical regeneration of granular activated carbon saturated with organic compounds. *Chem. Eng. J.* **2009**, *155*, 763–768. [CrossRef]

6. Asghar, H.M.A.; Hussain, S.N.; Roberts, E.P.L.; Brown, N.W. Removal of humic acid from water using adsorption coupled with electrochemical regeneration. *Korean J. Chem. Eng.* **2013**, *30*, 1415–1422. [CrossRef]
7. Asghar, H.M.A.; Hussain, S.N.; Sattar, H.; Brown, N.W.; Roberts, E.P.L. Electrochemically synthesized GIC-based adsorbents for water treatment through adsorption and electrochemical regeneration. *J. Ind. Eng. Chem.* **2014**, *20*, 2200–2207. [CrossRef]
8. Narbaitz, R.M.; Cen, J. Electrochemical regeneration of granular activated carbon. *Water Res.* **1994**, *28*, 1771–1778. [CrossRef]
9. García-Otón, M.; Montilla, F.; Lillo-Rodenas, M.A.; Morallon, E.; Vázquez, J.L. Electrochemical regeneration of activated carbon saturated with toluene. *J. Appl. Electrochem.* **2005**, *35*, 319–325. [CrossRef]
10. Han, Y.; Quan, X.; Ruan, X.; Zhang, W. Integrated electrochemically enhanced adsorption with electrochemical regeneration for removal of acid orange 7 using activated carbon fibers. *Sep. Purif. Technol.* **2008**, *59*, 43–49. [CrossRef]
11. Brown, N.; Roberts, E.P.; Garforth, A.; Dryfe, R.A. Electrochemical regeneration of a carbon-based adsorbent loaded with crystal violet dye. *Electrochim. Acta* **2004**, *49*, 3269–3281. [CrossRef]
12. Brown, N.W.; Roberts, E.P.L.; Chasiotis, A.; Cherdron, T.; Sanghrajka, N. Atrazine removal using adsorption and electrochemical regeneration. *Water Res.* **2004**, *38*, 3067–3074. [CrossRef]
13. Brown, N.W.; Roberts, E.P.L. Electrochemical pre-treatment of effluents containing chlorinated compounds using an adsorbent. *J. Appl. Electrochem.* **2007**, *37*, 1329–1335. [CrossRef]
14. Sharif, F.; Gagnon, L.R.; Mulmi, S.; Roberts, E.P.L. Electrochemical regeneration of a reduced graphene oxide/magnetite composite adsorbent loaded with methylene blue. *Water Res.* **2017**, *114*, 237–245. [CrossRef] [PubMed]
15. Nkrumah-Amoako, K.; Roberts, E.P.L.; Brown, N.W.; Holmes, S.M. The effects of anodic treatment on the surface chemistry of a Graphite Intercalation Compound. *Electrochim. Acta* **2014**, *135*, 568–577. [CrossRef]
16. Rueffer, M.; Bejan, D.; Bunce, N.J. Graphite: An active or an inactive anode? *Electrochim. Acta* **2011**, *56*, 2246–2253. [CrossRef]
17. Ntsendwana, B.; Mamba, B.B.; Sampath, S.; Arotiba, O.A. Synthesis, characterisation and application of an exfoliated graphite-diamond composite electrode in the electrochemical degradation of trichloroethylene. *RSC Adv.* **2013**, *3*, 24473–24483. [CrossRef]
18. Peleyeju, M.G.; Umukoro, E.H.; Babalola, J.O.; Arotiba, O.A. Electrochemical Degradation of an Anthraquinonic Dye on an Expanded Graphite-Diamond Composite Electrode. *Electrocatalysis* **2016**, *7*, 132–139. [CrossRef]
19. Yang, Y.; Hoffmann, M.R. Synthesis and Stabilization of Blue-Black TiO_2 Nanotube Arrays for Electrochemical Oxidant Generation and Wastewater Treatment. *Environ. Sci. Technol.* **2016**, *50*, 11888–11894. [CrossRef]
20. Jasmann, J.R.; Borch, T.; Sale, T.C.; Blotevogel, J. Advanced Electrochemical Oxidation of 1, 4-Dioxane via Dark Catalysis by Novel Titanium Dioxide (TiO2) Pellets. *Environ. Sci. Technol.* **2016**, *50*, 8817–8826. [CrossRef]
21. Li, X.; Wu, Y.; Zhu, W.; Xue, F.; Qian, Y.; Wang, C. Enhanced electrochemical oxidation of synthetic dyeing wastewater using SnO_2-Sb-doped TiO_2-coated granular activated carbon electrodes with high hydroxyl radical yields. *Electrochim. Acta* **2016**, *220*, 276–284. [CrossRef]
22. Guo, X.; Li, D.; Wan, J.; Yu, X. Preparation and electrochemical property of TiO_2/Nano-graphite composite anode for electro-catalytic degradation of ceftriaxone sodium. *Electrochim. Acta* **2015**, *180*, 957–964. [CrossRef]
23. Wang, L.; Zhao, Y.; Fu, J. The influence of TiO_2 and aeration on the kinetics of electrochemical oxidation of phenol in packed bed reactor. *J. Hazard. Mater.* **2008**, *160*, 608–613. [CrossRef] [PubMed]
24. Cui, X.; Zhao, G.; Lei, Y.; Li, H.; Li, P.; Liu, M. Novel vertically aligned TiO_2 nanotubes embedded with Sb-doped SnO_2 electrode with high oxygen evolution potential and long service time. *Mater. Chem. Phys.* **2009**, *113*, 314–321. [CrossRef]
25. Xie, R.; Meng, X.; Sun, P.; Niu, J.; Jiang, W.; Bottomley, L.; Li, D.; Chen, Y.; Crittenden, J. Electrochemical oxidation of ofloxacin using a TiO_2-based SnO_2-Sb/polytetrafluoroethylene resin-PbO_2 electrode: Reaction kinetics and mass transfer impact. *Appl. Catal. B Environ.* **2017**, *203*, 515–525. [CrossRef]
26. Beck, F.; Kaiser, W.; Krohn, H. Boron doped diamond (BDD)-layers on titanium substrates as electrodes in applied electrochemistry. *Electrochim. Acta* **2000**, *45*, 4691–4695. [CrossRef]

27. Song, S.; Fan, J.; He, Z.; Zhan, L.; Liu, Z.; Chen, J.; Xu, X. Electrochemical degradation of azo dye CI Reactive Red 195 by anodic oxidation on Ti/SnO_2–Sb/PbO_2 electrodes. *Electrochim. Acta* **2010**, *55*, 3606–3613. [CrossRef]
28. Siedlecka, E.M.; Fabianska, A.; Stolte, S.; Nienstedt, A.; Ossowski, T.; Stepnowski, P.; Thöming, J. Electrocatalytic oxidation of 1-butyl-3-methylimidazolium chloride: Effect of the electrode material. *Int. J. Electrochem. Sci.* **2013**, *8*, 5560–5574.
29. Sharif, F.; Roberts, E.P.L. Anodic electrochemical regeneration of a graphene/titanium dioxide composite adsorbent loaded with an organic dye. *Chemosphere* **2019**, *241*, 125020. [CrossRef]
30. Zhao, G.; Wen, T.; Yang, X.; Yang, S.; Liao, J.; Hu, J.; Shao, D.; Wang, X. Preconcentration of U (VI) ions on few-layered graphene oxide nanosheets from aqueous solutions. *Dalton Trans.* **2012**, *41*, 6182–6188. [CrossRef]
31. Zuo, J.; Xu, C.; Liu, X.; Wang, C.; Wang, C.; Hu, Y.; Qian, Y. Study of the Raman spectrum of nanometer SnO_2. *J. Appl. Phys.* **1994**, *75*, 1835–1836. [CrossRef]
32. Camacho-López, M.A.; Galeana-Camacho, J.R.; Esparza-García, A.; Sánchez-Pérez, C.; Julien, C.M. Characterization of nanostructured SnO_2 films deposited by reactive DC-magnetron sputtering. *Superf. Vacío* **2013**, *26*, 95–99.
33. Xu, J.M.; Li, L.; Wang, S.; Ding, H.L.; Zhang, Y.X.; Li, G.H. Influence of Sb doping on the structural and optical properties of tin oxide nanocrystals. *CrystEngComm* **2013**, *15*, 3296–3300. [CrossRef]
34. Bonu, V.; Das, A.; Sivadasan, A.K.; Tyagi, A.K.; Dhara, S. Invoking forbidden modes in SnO_2 nanoparticles using tip enhanced Raman spectroscopy. *J. Raman Spectrosc.* **2015**, *46*, 1037–1040. [CrossRef]
35. Velmurugan, V.; Srinivasarao, U.; Ramachandran, R.; Saranya, M.; Grace, A.N. Synthesis of tin oxide/graphene (SnO_2/G) nanocomposite and its electrochemical properties for supercapacitor applications. *Mater. Res. Bull.* **2016**, *84*, 145–151. [CrossRef]
36. Hameed, B.H.; Rahman, A.A. Removal of phenol from aqueous solutions by adsorption onto activated carbon prepared from biomass material. *J. Hazard. Mater.* **2008**, *160*, 576–581. [CrossRef]
37. De la Luz-Asunción, M.; Sánchez-Mendieta, V.; Martínez-Hernández, A.L.; Castaño, V.M.; Velasco-Santos, C. Adsorption of phenol from aqueous solutions by carbon nanomaterials of one and two dimensions: Kinetic and equilibrium studies. *J. Nanomater.* **2015**, *16*, 422. [CrossRef]
38. Calcaterra, D.; Parise, M. Weathering as a predisposing factor to slope movements: An introduction. *Geol. Soc. Lond. Eng. Geol. Spec. Publ.* **2010**, *23*, 1–4. [CrossRef]
39. Lau, C.H.; Chen, S.F. Methylene blue adsorption by clays saturated with different inorganic cations. *Int. Inf. Syst. Agric. Sci. Technol.* **1978**, *26*, 59–66.
40. Zhang, H. Regeneration of exhausted activated carbon by electrochemical method. *Chem. Eng. J.* **2002**, *85*, 81–85. [CrossRef]
41. Asghar, H.M.A.; Hussain, S.N.; Roberts, E.P.L.; Campen, A.K.; Brown, N.W. Pre-treatment of adsorbents for waste water treatment using adsorption coupled-with electrochemical regeneration. *J. Ind. Eng. Chem.* **2013**, *19*, 1689–1696. [CrossRef]
42. Radjenovic, J.; Sedlak, D.L. Challenges and opportunities for electrochemical processes as next-generation technologies for the treatment of contaminated water. *Environ. Sci. Technol.* **2015**, *49*, 11292–11302. [CrossRef] [PubMed]
43. Zhao, G.; Cui, X.; Liu, M.; Li, P.; Zhang, Y.; Cao, T.; Li, H.; Lei, Y.; Liu, L.; Li, D. Electrochemical Degradation of Refractory Pollutant Using a Novel Microstructured TiO_2 Nanotubes/Sb-Doped SnO_2 Electrode. *Environ. Sci. Technol.* **2009**, *43*, 1480–1486. [CrossRef] [PubMed]
44. Wang, Q.; Jin, T.; Hu, Z.; Zhou, L.; Zhou, M. TiO_2-NTs/SnO_2-Sb anode for efficient electrocatalytic degradation of organic pollutants: Effect of TiO_2-NTs architecture. *Sep. Purif. Technol.* **2013**, *102*, 180–186. [CrossRef]
45. Nurzulaikha, R.; Lim, H.N.; Harrison, I.; Lim, S.S.; Pandikumar, A.; Huang, N.M.; Lim, S.P.; Thien, G.S.H.; Yusoff, N.; Ibrahim, I. Graphene/SnO_2 nanocomposite-modified electrode for electrochemical detection of dopamine. *Sens. Bio-Sens. Res.* **2015**, *5*, 42–49. [CrossRef]
46. Yao, J.; Shen, X.; Wang, B.; Liu, H.; Wang, G. In situ chemical synthesis of SnO_2–graphene nanocomposite as anode materials for lithium-ion batteries. *Electrochem. Commun.* **2009**, *11*, 1849–1852. [CrossRef]
47. Kong, J.; Shi, S.; Zhu, X.; Ni, J. Effect of Sb dopant amount on the structure and electrocatalytic capability of Ti/Sb-SnO_2 electrodes in the oxidation of 4-chlorophenol. *J. Environ. Sci.* **2007**, *19*, 1380–1386. [CrossRef]

48. He, D.; Mho, S. Electrocatalytic reactions of phenolic compounds at ferric ion co-doped SnO_2: Sb5+ electrodes. *J. Electroanal. Chem.* **2004**, *568*, 19–27. [CrossRef]
49. Caldararu, M.; Thomas, M.F.; Bland, J.; Spranceana, D. Redox processes in Sb-containing mixed oxides used in oxidation catalysis: I. Tin dioxide assisted antimony oxidation in solid state. *Appl. Catal. A Gen.* **2001**, *209*, 383–390. [CrossRef]
50. Gao, J.; Qin, Y.; Zhou, T.; Cao, D.; Xu, P.; Hochstetter, D.; Wang, Y. Adsorption of methylene blue onto activated carbon produced from tea (*Camellia sinensis* L.) seed shells: Kinetics, equilibrium, and thermodynamics studies. *J. Zhejiang Univ. Sci. B* **2013**, *14*, 650–658. [CrossRef]
51. Hussain, S.N.; Asghar, H.M.A.; Sattar, H.; Brown, N.W.; Roberts, E.P.L. Removal of Tartrazine From Water by Adsorption with Electrochemical Regeneration. *Chem. Eng. Commun.* **2015**, *202*, 1280–1288. [CrossRef]
52. Mohammed, F.M.; Roberts, E.P.L.; Campen, A.K.; Brown, N.W. Wastewater treatment by multi-stage batch adsorption and electrochemical regeneration. *J. Electrochem. Sci. Eng.* **2012**, *2*, 223–236. [CrossRef]

Article

Fenton and Photo-Fenton Nanocatalysts Revisited from the Perspective of Life Cycle Assessment

Sara Feijoo [1], Jorge González-Rodríguez [1], Lucía Fernández [1], Carlos Vázquez-Vázquez [2], Gumersindo Feijoo [1] and María Teresa Moreira [1,*]

1. CRETUS Institute, Department of Chemical Engineering, Universidade de Santiago de Compostela, 15782 Santiago de Compostela, Spain; sarafeijoomoreira@yahoo.es (S.F.); jorgegonzalez.rodriguez@usc.es (J.G.-R.); lucia.fernandezf@gmail.com (L.F.); gumersindo.feijoo@usc.es (G.F.)
2. Laboratory of Magnetism and Nanotechnology, Department of Physical Chemistry, Faculty of Chemistry, Universidade de Santiago de Compostela, 15782 Santiago de Compostela, Spain; carlos.vazquez.vazquez@usc.es

* Correspondence: maite.moreira@usc.es; Tel.: +34-8818-16792

Received: 8 November 2019; Accepted: 22 December 2019; Published: 24 December 2019

Abstract: This study provides an overview of the environmental impacts associated with the production of different magnetic nanoparticles (NPs) based on magnetite (Fe_3O_4), with a potential use as heterogeneous Fenton or photo-Fenton catalysts in wastewater treatment applications. The tendency of Fe_3O_4 NPs to form aggregates in water makes necessary their decoration with stabilizing agents, in order to increase their catalytic activity. Different stabilizing agents were considered in this study: poly(acrylic acid) (PAA), polyethylenimine (PEI) and silica (SiO_2), as well as the immobilization of the magnetite-based catalysts in a mesoporous silica matrix, SBA-15. In the case of photo-Fenton catalysts, combinations of magnetite NPs with semiconductors were evaluated, so that magnetic recovery of the nanomaterials is possible, thus allowing a safe discharge free of NPs. The results of this study suggest that magnetic nanoparticles coated with PEI or PAA were the most suitable option for their applications in heterogeneous Fenton processes, while ZnO-Fe_3O_4 NPs provided an interesting approach in photo-Fenton. This work showed the importance of identifying the relevance of nanoparticle production strategy in the environmental impacts associated with their use.

Keywords: LCA; environmental impact; Fenton; photocatalysis; visible light; SBA15; magnetite

---

1. Introduction

In the era of nanotechnology, there is a growing interest in the use of nanomaterials, which is largely attributed to their characteristic high specific surface and reactivity. In this regard, nanomaterials are emerging as an interesting alternative in multiple applications in substitution of bulk chemicals, from biomedical and clinical diagnosis to their use in the field of biochemical catalysis and environmental engineering. Advanced oxidation processes (AOPs) based on the use of NPs have shown great potential for the treatment of industrial wastewaters [1–5]. These processes encompass the generation of highly reactive oxygen species (ROS), such as hydroxyl radicals ($HO^{\bullet}$), superoxide radicals ($O_2^{\bullet-}$) and singlet oxygen ($^1O_2$), which are involved in the degradation of organic matter, generally leading to more biodegradable species with less environmental impact.

One of the most widely studied AOP processes is the Fenton process, which is based on the use of iron as a catalyst to improve the oxidative potential of H_2O_2. This reaction takes place at an acid pH of around 3 [6], which favors the formation of radicals. Magnetite (Fe_3O_4) is a naturally occurring mixed iron oxide constituted by Fe(II) and Fe(III). While bulk Fe_3O_4 is ferromagnetic, Fe_3O_4 NPs are superparamagnetic, which provides them with a stronger magnetic response when exposed to a magnetic field. This feature makes Fe_3O_4 NPs of special interest in comparison with traditional iron-based catalysts, since their magnetic nature allows easy, fast and economical separation of the NPs from the reaction medium, which facilitates their reuse in subsequent catalytic cycles [7].

Fe_3O_4 NPs have also been widely applied as supports for photocatalytic agents such as zinc oxide (ZnO) or titanium dioxide (TiO_2), which show promising catalytic and photochemical properties towards the removal of persistent organic pollutants, such as pharmaceutical and personal care products (PPCPs) and dyes [8]. The immobilization of TiO_2 or ZnO NPs in Fe_3O_4 NPs provides not only easy recovery of the photocatalyst, but also a dual-nature nanocomposite as a photo-Fenton agent.

However, there is growing concern regarding possible environmental impacts associated not only with the use of nanoparticles, but also with the synthesis phase. The fact that they provide more reactivity than the original compounds has been the driving force of research aimed at assessing their potential toxicity in terms of human health and ecosystem damage. The toxicity of nanoparticles depends on multiple factors, such as size, crystalline structure, surface coatings, or the presence of co-pollutants in the media, characteristics that may facilitate their transport through cell membranes. However, the magnetic properties of the nanoparticles allow their retention in wastewater treatment processes, preventing their release into the environment [9,10]. These two objective indicators can guide the efforts needed for risk assessment. In addition, the production process can also lead to impacts on different environmental compartments, which must be identified in order to identify the critical aspects that affect all stages of the life cycle, not only the final product, but also the resources (chemical, energy, water) and the emissions directly and indirectly associated with their synthesis and stabilization.

Fe_3O_4 NPs can be prepared through different chemical approaches, such as co-precipitation, microemulsion, sol-gel, sonochemical and thermal decomposition, solvothermal, microwave-assisted, chemical vapor deposition, and laser pyrolysis [11–15]. Among these, chemical co-precipitation is probably the most common approach, and is based on the reaction of an aqueous mixture of ferrous and ferric salts in alkaline medium that renders into Fe_3O_4 NPs with a broad size distribution, ranging from 4 to 20 nm. To produce NPs with a more homogeneous size distribution, water-in-oil microemulsion can be considered, which consists of nano-sized water droplets dispersed in an oil phase, stabilized by a surfactant at the water/oil interface. These nano-cavities are typically in the range of 10 nm and provide a confinement effect that limits the particle nucleation, growth and agglomeration [16].

On the other hand, Fe_3O_4 NPs have a strong tendency to form aggregates in water due to their high surface energy and van der Waals interactions [17]. To avoid this hindrance, Fe_3O_4 NPs can be decorated with protecting agents, including organic coatings such as polymers (polyethylene glycol, polyvinylpyrrolidone, polyvinyl alcohol, polystyrene), natural products (chitosan, gelatin, starch, ethyl cellulose), surfactants (oleic acid, lauric acid), and inorganic coatings such as metal oxides, SiO_2 or carbon [18]. However, an increase in the stability of Fe_3O_4 in water involves difficulty in recovering them and, in the worst-case scenario, an additional separation stage between water treatment cycles. An alternative of growing interest is the use of mesoporous material, used as a support for active chemical agents due to its large surface area, mechanical and chemical stability.

Considering an integrated approach of the Fenton process and photocatalysis, the combination of a semiconductor such as TiO_2 or ZnO with a magnetic nanocomposite allows the effective separation of the reaction medium through the application of a magnetic field [19]. From the point of view of chemical synthesis, the process presents the complexity of integrating both routes, and the efficiency of the nanocatalyst can be affected in relation to its properties, such as surface, adsorption, reflectance, adhesion, and transport properties.

When it comes to assessing the environmental impacts of NP production and use, the Life Cycle Assessment (LCA) methodology arises as the most updated and comprehensive approach [20], based on international standards (ISO 14040/44) [21]. Available LCA studies of nanomaterials include nanomaterials such as CdTe, carbon nanofibres (CNFs) and nanotubes (CNTs), nanoclay (ONMT, organically modified montmorillonite), nanoscale Pt-group metals, nanocrystaline-Si, Ag, titanium and titanium oxide, and magnetic NPs [22–25]. However, to date, no LCA reports have been found on NP production routes for Fenton and photo-Fenton applications.

The aim of this work is to benchmark the different nanostructured materials used as Fenton and photo-Fenton catalysts. The synthesis of Fe_3O_4 NPs used as Fenton agents will consider their steric or electrostatic stabilization, as well as their immobilization in SBA-15 mesoporous matrix. The synthesis of hybrid nanocomposites of TiO_2 or ZnO NPs supported onto Fe_3O_4 NPs will also be considered. The nanocatalysts with the most benign environmental profiles from the different groups will be selected on the basis of the environmental indicators associated with the production process as well as their catalytic capacity for oxidation of different dyes, chosen as model compounds of organic pollutants. Both analyses will provide additional information that will serve as criteria in the decision-making process to identify the most suitable AOP nanocatalyst.

2. Results

The proposal of any process under development which aims at replacing more developed alternatives must clearly demonstrate the expected benefits, not only from the point of view of technological feasibility and cost, but also of the environmental impacts associated with the process to be developed. What is the point if the global analysis shows that beyond the capability of certain NPs to degrade target chemicals with high performance, we are introducing higher environmental loads associated with the production of the nanocatalyst?

The assessment of the environmental impacts associated with the production of NPs was based on inventory data obtained on a pilot scale. In this case, it should be noted that all the data related to the production of nanoparticles correspond to real data, where the production protocol is presented in the materials and method section. Tables 1 and 2 present the Life Cycle Impact Assessment results for the different production routes of Fenton catalysts and photocatalysts. The different routes in terms of impact categories highlight that the synthesis protocols for the different NP present a great variability in the results obtained, identifying that the use of energy and chemical products during NP synthesis presents a significant and often dominant part of the environmental impacts of the total life cycle.

Table 1. Life Cycle Impact Assessment results for different types of Fenton catalysts.

	Fe_3O_4	Fe_3O_4-SiO_2	Fe_3O_4-SiO_2SBA15	Fe_3O_4-PEI	Fe_3O_4-PEISBA15	Fe_3O_4-PAA	Fe_3O_4-PAA SBA15
Climate change (CC), kg $CO_{2,eq}$	$5.33 \cdot 10^{-2}$	1.86	20.3	$1.49 \cdot 10^{-1}$	6.94	$7.90 \cdot 10^{-2}$	6.94
Ozone depletion (OD), kg CFC-11_{eq}	$8.21 \cdot 10^{-9}$	$1.59 \cdot 10^{-7}$	$2.67 \cdot 10^{-6}$	$2.05 \cdot 10^{-8}$	$8.64 \cdot 10^{-7}$	$1.29 \cdot 10^{-8}$	$8.64 \cdot 10^{-7}$
Terrestrial acidification (TA), kg $SO_{2,eq}$	$3.26 \cdot 10^{-4}$	$8.75 \cdot 10^{-3}$	$1.20 \cdot 10^{-1}$	$8.86 \cdot 10^{-4}$	$3.95 \cdot 10^{-2}$	$4.79 \cdot 10^{-4}$	$3.95 \cdot 10^{-2}$
Freshwater eutrophication (FE), kg P_{eq}	$1.37 \cdot 10^{-5}$	$4.35 \cdot 10^{-4}$	$6.30 \cdot 10^{-3}$	$4.54 \cdot 10^{-5}$	$2.14 \cdot 10^{-3}$	$2.46 \cdot 10^{-5}$	$2.14 \cdot 10^{-3}$
Marine eutrophication (ME), kg N_{eq}	$1.12 \cdot 10^{-5}$	$2.97 \cdot 10^{-4}$	$4.11 \cdot 10^{-3}$	$3.91 \cdot 10^{-5}$	$1.43 \cdot 10^{-3}$	$1.75 \cdot 10^{-5}$	$1.42 \cdot 10^{-3}$
Toxicity (TOX), kg 1,4-DCB_{eq}	$1.90 \cdot 10^{-2}$	$5.32 \cdot 10^{-1}$	7.32	$5.53 \cdot 10^{-2}$	2.38	$3.18 \cdot 10^{-2}$	2.38
Fossil depletion (FD), kg oil_{eq}	$1.67 \cdot 10^{-2}$	1.01	5.82	$3.98 \cdot 10^{-2}$	1.99	$2.34 \cdot 10^{-2}$	1.99

Table 2. Life Cycle Impact Assessment results for different photocatalysts.

	Fe_3O_4/ZnO	Fe_3O_4/TiO_2
Climate change (CC), kg $CO_{2,eq}$	1.78	6.08
Ozone depletion (OD), kg CFC-11_{eq}	$1.66 \cdot 10^{-7}$	$8.30 \cdot 10^{-7}$
Terrestrial acidification (TA), kg $SO_{2,eq}$	$7.99 \cdot 10^{-3}$	$3.77 \cdot 10^{-2}$
Freshwater eutrophication (FE), kg P_{eq}	$5.91 \cdot 10^{-4}$	$1.88 \cdot 10^{-3}$
Marine eutrophication (ME), kg N_{eq}	$2.91 \cdot 10^{-4}$	$1.31 \cdot 10^{-3}$
Toxicity (TOX), kg 1,4-DCB_{eq}	$5.25 \cdot 10^{-1}$	2.29
Fossil depletion (FD), kg oil_{eq}	$6.00 \cdot 10^{-1}$	1.71

When analyzing the environmental profile of NPs considered to be Fenton catalysts, two distinct groups are observed: the stabilized magnetite NPs and those with a surface coating (Group 1); and those that have been immobilized in a mesoporous matrix (Group 2). It is evident that there are notable differences between the NPs belonging to one group and those belonging to the other, so that the environmental impacts associated with mesoporous NPs are excessively high in comparison with those produced with a simpler procedure. Accordingly, the high environmental impact of the Fe_3O_4 NPs onto the SBA-15 matrix can be explained attending to the use of additional chemicals needed to produce the mesoporous silica, as well as the electricity used during the heating stages required to carry out the reaction.

When analyzing in detail the different routes, it is also evident that some of the alternatives evaluated resemble each other due to the comparable production of nanoparticles, similar use of chemicals and electricity, and minor differences in the amount of chemicals used to adjust pH, water consumption for the washing stages, and similar wastewater flows generated. These similarities are especially evident for the production of Fe_3O_4, Fe_3O_4@PEI and Fe_3O_4@PAA NPs, which present similar environmental profiles and limited impacts, clearly lower than those of silica-coated NPs with a more complex production scheme (Figure 1). Beyond the stages common to the different pathways, the main differences are related to the complexity of the downstream stages necessary for the final formulation of the different nanoparticles. On the other hand, during the production of Fe_3O_4@SiO_2 NPs, cyclohexane is used in the re-dispersion and in the reaction stages, which represents more than 25% of the impact of chemicals in the manufacturing process. From an environmental point of view, the use of PEI or PAA coated NPs would be recommended, even preferentially to the sterically stabilized ones. Although the latter have less impact, they also tend to agglomerate in a shorter period of time, so that their operational stability may be compromised.

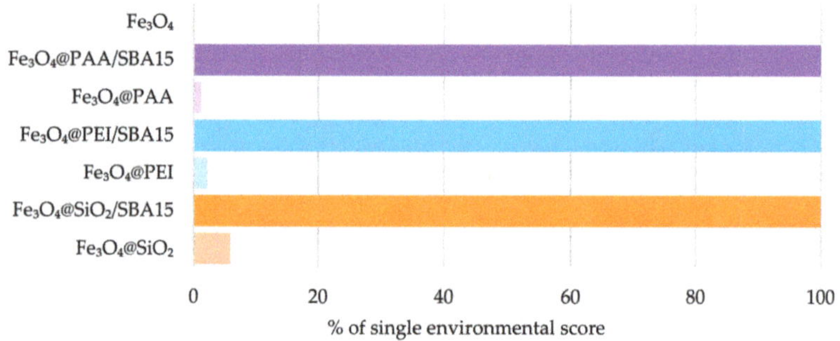

Figure 1. Benchmarking relative values of single environmental scores for the synthesis routes of magnetic NPs.

When reporting the results of nanoparticles used for photocatalysis, the environmental impact of Fe_3O_4/ZnO production show values comparable to those of silica-coated NPs, but there is a clear difference when assessing the impacts of Fe_3O_4/TiO_2 nanocomposite, which implies a substantial increase in the normalized environmental impact up to $7.1 \cdot 10^{-2}$ (Figure 2).

Beyond the environmental impacts of their production, the focus should be on considering these indicators and the oxidation potential of nanocatalysts towards a target compound. For this purpose, a selection of those nanocatalysts with promising results from each group were evaluated for the removal of Reactive Blue 19 (RB19), Congo Red (CR), Orange II (OII) and Methyl Green (MG), which were selected as models of organic pollutants.

Degradation depends on multiple factors, such as the interaction between catalysts and compounds, the stability of the catalyst, the size of the nanoparticles or the specific surface. The surface charge of the nanoparticles studied was measured: only Fe_3O_4@PEI was positively charged, with a zeta potential of 29.9 mV. The other Fenton catalysts had surface loads of −11.5, −14.0 and −14.3 mV for

Fe_3O_4, Fe_3O_4@PAA, and Fe_3O_4@SiO_2, respectively (Figure 3). The size distribution of the magnetite nanoparticles was about 10 nm, and in the case of the silica-coated catalyst, the thickness of the shell was about 5 nm. More specific data on the characterization of nanoparticles can be found in the list of references corresponding to the methodology section.

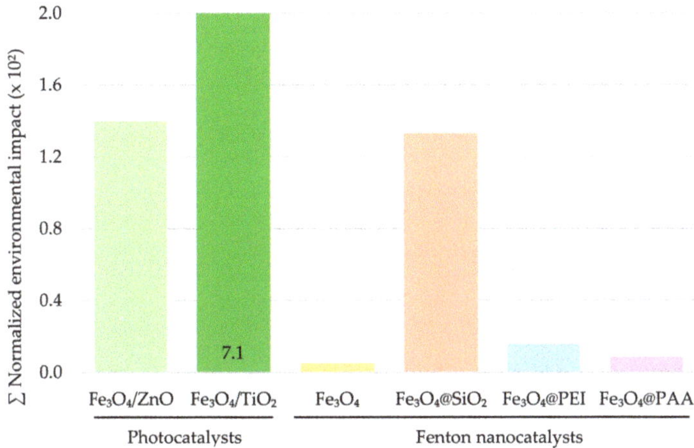

Figure 2. Environmental impacts associated with the synthesis routes of magnetic NPs.

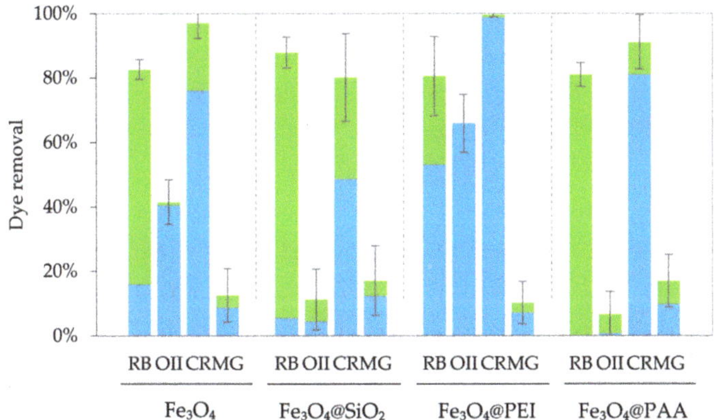

Figure 3. Percentage of dye adsorption (blue) and degradation (green) for $C_0 = 10$ mg L^{-1} after 3 h of Fenton (300 mg L^{-1} of catalyst, 300 mg L^{-1} of H_2O_2, pH 3).

Depending on the type of dye, the decolorization results are different: the RB19 dye is degraded more extensively than the MG dye, whose degradation is in the best-case scenario lower than 20%. Surprisingly, the performance of the Fe_3O_4/ZnO photocatalyst was superior to that of the Fenton catalysts, reaching elimination levels close to 100%. However, for the other dyes, its performance was clearly lower, with removal percentages below 50% (data not shown). As for the Fenton catalysts, the performance of the different alternatives in terms of dye removal was quite similar in all cases, but the stability of the catalysts improved with the addition of coatings. It was observed that the coating of nanoparticles with PEI substantially improved the adsorption of the target compounds due to their surface load, but the degradation values for this nanocatalyst were very low. The results also show the influence of the type of the target compound on degradation, related to the surface charge and the interaction between the dye and the nanoparticle.

The reuse of nanoparticles is interesting for wastewater treatment applications, as the costs associated with catalyst production are reduced. Selected nanoparticles include bare magnetite nanoparticles, as a basis for comparison, and other nanoparticles with different coatings. On the one hand, organic coatings with PAA and PEI provide stabilization, because the repulsive forces between these chemically bonded polymers and the nanoparticles reduce the super-paramagnetic interactions of the nanostructured magnetite. On the other hand, the coating with SiO_2 provides stability by forming a layer around the magnetic nanoparticles. In addition, the immobilization of semiconductor photocatalysts on magnetic supports allows separation by applying a magnetic field, improving recovery.

Considering the combination of technical and environmental perspectives, although the nanocomposite PAA-coated Fe_3O_4/SBA15 could be preferred among the different alternatives, the environmental impact was significantly greater for this option. Considering a balance between efficiency and impact results, a preferred option is the option of PAA-coated Fe_3O_4 as it combines catalytic potential, simplicity in production, magnetic separation of the catalyst and low environmental impact.

3. Materials and Methods

3.1. Reagents

Tetraethyl orthosilicate (TEOS, 98% *w*/*w*), iron(III) chloride hexahydrate ($FeCl_3 \cdot 6H_2O$, 97% w/w), zinc(II) acetate dihydrate ($Zn(CH_3COO)_2 \cdot 2H_2O$, ≥98% w/w), titanium(IV) isopropoxide ($Ti(OCH(CH_3)_2)_4$, ≥97% *w*/*w*), hydrochloric acid (HCl, 37% *w*/*w*), cyclohexane (C_6H_{12}, 99.8% *w*/*w*), Igepal CO-520 (polyoxyethylene (5) nonylphenylether, branched), polyacrylic acid (PAA, Mw 2000), polyethylenimine (PEI, Mw 25000), triblock copolymer Pluronic P123 (PEO20-PPO70-PEO20), 2-propanol ($(CH_3)_2CHOH$, ≥99.5% *w*/*w*), Reactive Blue 19 (RB19, ≈50% *w*/*w*), Congo Red (CR, ≥85% *w*/*w*), Orange II (OII, ≥85% *w*/*w*), Methyl Green (MG, 85% *w*/*w*) and hydrogen peroxide (H_2O_2, 30% *w*/*w*) were obtained from Sigma-Aldrich (St. Louis, MI, USA). Iron(II) sulfate heptahydrate ($FeSO_4 \cdot 7H_2O$, 99% *w*/*w*), ortho phosphoric acid (H_3PO_4, 85% *w*/*w*), tetramethylammonium hydroxide (TMAOH, ≈10% *w*/*w*) and ammonium hydroxide (NH_4OH, 28% *w*/*w*) from Fluka (Buchs, ZU, Switzerland); ethanol (CH_3CH_2OH, 99.9% *w*/*w*) and acetone (CH_3COCH_3, ≥99% *w*/*w*) were purchased from Scharlau (Sentmenat, CT, Spain), and oleic acid (cis-9-octadecenoic acid, $CH_3(CH_2)_7CH{=}CH(CH_2)_7COOH$, extra pure) from Merck (Darmstadt, HE, Germany). Milli-Q deionized water was used in all the experiments and were purchased from Merck-Millipore (Darmstadt, HE, Germany).

3.2. Goal and Scope Definition

The purpose of this work is to evaluate the environmental performance of different types of nanoparticles as Fenton and photo-Fenton catalysts based on the use of magnetite NPs stabilized with surfactants or coatings, immobilized on mesostructured materials or combined with semiconductors. The selected nanoparticles are listed in Table 3.

Table 3. Selected nanoparticles used in different AOPs description.

AOP Type	Nanocatalyst Code	Nanocatalyst Description
Fenton process	Fe_3O_4	Magnetite NPs
	Fe_3O_4@PAA	Poly(acrylic acid) coated magnetic NPs
	Fe_3O_4@PEI	Polyethylenimine coated magnetic NPs
	Fe_3O_4@SiO_2	Silica coated magnetic NPs
	Fe_3O_4@PAA/SBA15	Poly(acrylic acid) coated magnetic NPs supported onto SBA-15
	Fe_3O_4@PEI/SBA15	Polyethylenimine coated magnetic NPs supported onto SBA-15
	Fe_3O_4@SiO_2/SBA15	Silica coated magnetic NPs supported onto SBA-15
Photocatalysis process	Fe_3O_4/TiO_2	Titanium dioxide magnetic nanocomposite
	Fe_3O_4/ZnO	Zinc oxide magnetic nanocomposite

3.3. *Description of the Different Production Processes*

3.3.1. Sterically Stabilized Magnetite

In a typical synthesis, $FeCl_3$ (45 mmol) and $FeSO_4$ (Fe^{3+}/Fe^{2+} molar ratio ≈ 1.5) are dissolved in a 100 mL of 0.01 M HCl solution. Temperature increased to 60 °C in a 250 mL round-bottom flask, using 30 mL of NH_4OH (28% *w/w*) in a subsequent step to promote the formation of black magnetite nanoparticles. The nanoparticles were washed three times with deionized water and, finally, a tetramethylammonium hydroxide (TMAOH) solution (10% *w/w*) was added to reach pH 10 to obtain sterically stabilized magnetite [25]. Both the energy consumption associated with agitation and heating (60 °C) and the consumption of deionized water for redispersion and washing were considered in the computation of inventory data. Oleic acid (OA) is a surfactant commonly used to stabilize black magnetite nanoparticles, although it can be considered as an intermediate step in the formulation of more complex NPs, such as those coated with silica. For this purpose, Fe_3O_4 NPs are exposed to an OA solution until flocculation occurs.

3.3.2. Mesostructured Silica Synthesis

The procedure for producing mesoporous ordered silica SBA-15 was based on the Colilla method [26], using a Pluronic P123 (PEO20-PPO70-PEO20) triblock copolymer, HCl/H_3PO_4 and TEOS. The product was dried and subjected to different washing cycles with organic solvents to remove the remaining block copolymer [27]. For the life cycle inventory (LCI) estimation, the energy consumption associated with agitation (24 h), heating (35–100 °C) and drying (60 °C) and the consumption of isopropyl alcohol for washing were considered.

3.3.3. Synthesis of Iron Oxide Supported in Mesostructured Silica

The addition of SBA-15 to the solvothermal preparation of Fe_3O_4 NPs allowed to obtain multi-core magnetic mesoporous nanocomposites [27]. As in previous protocols, the energy consumption associated with mechanical stirring and drying (60 °C) and the consumption of ethanol and distilled water for washing were considered.

3.3.4. Polyethylenimine (PEI)-Coated and Polyacrylic Acid (PAA)-Coated Magnetite Nanoparticles

The production process starts with the addition of PEI or PAA to Fe_3O_4 NPs, with the addition of HCl to lower the pH to 4 before the addition of tetraethyl orthosilicate (TEOS) to produce mNP [28]. For the LCI estimation, the energy consumption associated with mechanical stirring and drying (60 °C) and the consumption of ethanol and distilled water for washing were considered.

3.3.5. Silica-Coated Magnetic Nanoparticles

$Fe_3O_4@SiO_2$ core-shell nanoparticles were prepared through a water-in-cyclohexane reverse microemulsion starting from the oleic-acid-coated magnetite nanoparticles [25], using as main chemicals: polyoxyethylene(5)nonylphenyl ether (Igepal CO-520), cyclohexane, NH_4OH, TEOS and isopropanol (IPA) [29]. Beyond the consumption of chemicals and the energy consumption associated with mechanical stirring and drying (60 °C), the consumption of distilled water and cyclohexane for washing and redispersion were also included.

3.3.6. Preparation of Nanocomposites (NC) of Fe_3O_4@PAA/SBA-15 and Fe_3O_4@PEI/SBA-15

The nanocomposites were synthesized by incorporating the SBA-15 matrix in a container with an aqueous solution of HCl, $FeCl_3{\cdot}6H_2O$ and $Fe(SO_4)_2{\cdot}4H_2O$ under mechanical agitation. The temperature increased to 60 °C and NH_4OH and PAA (Mw 2000 Da) (Fe_3O_4@PAA/SBA-15) or PEI (Mw 25000 Da) were added to the mixture. The reaction was allowed to progress for 1 h, and the resulting precipitate was acidified to pH 4 with HCl (9%), and then magnetically separated. Finally, the NC were repeatedly washed with distilled water and ethanol and dried at 60 °C for 12 h [28]. For the LCI estimation, the energy consumption associated with mechanical stirring and drying (60 °C) and the consumption of ethanol and distilled water for washing were considered.

3.3.7. Preparation of the Nanoparticles Fe_3O_4@SiO_2/SBA-15

This procedure was based on the formation of silica-coated NPs in water-in-oil microemulsion systems, but including the SBA-15 matrix to favor the anchoring of Fe_3O_4@SiO_2 on the external surface of the SBA-15 matrix [29]. Finally, the reaction was continued at room temperature for 16 h, under mechanical stirring and in the dark. After the reaction was completed, isopropyl alcohol (IPA) was added to precipitate the material; the supernatant was separated from the magnetic solid with the help of a magnet. The sample was further washed several times with ethanol and deionized H_2O. Finally, the sample was filtered and dried at 60 °C for 24 h. In this case, the consumption of energy for mechanical stirring comprised several production phases: stage I (15 min), stage II (20 min), stage III (16 h) and drying (60 °C).

3.3.8. Synthesis of Fe_3O_4/ZnO Nanocomposite

The synthesis of Fe_3O_4/ZnO nanocomposite was based on the preparation of polyol-mediated ZnO nanoparticles [30], using 10 mL of diethyleneglycol (DEG) mixed with 100 mL of 90 mM Zn(II) acetate, in combination with 50 mg of sterically stabilized magnetite (TMAOH) obtained in previous steps. The mixture was heated at 180 °C for 2 h under mechanical agitation, and the as-prepared hybrid nanoparticles were centrifuged at 7500 rpm for 15 min. Then, they were washed four times with ethanol and magnetically separated. Finally, the hybrid nanoparticles were redispersed into ethanol at a concentration of approximately 1.5% determined by thermogravimetric analysis by Perkin Elmer TGA 7 (Waltham, MA, USA). For the LCI estimation, the energy consumption associated with mechanical stirring, heating (180 °C) and centrifugation and the consumption of ethanol for washing and redispersion were considered.

3.3.9. Synthesis of Fe_3O_4/TiO_2 Nanocomposite

The synthesis of Fe_3O_4@TiO_2 was based on the hydrolysis and polycondensation of titanium alkoxide, added dropwise to a mixture of TMAOH and 2-propanol [31]. After 10 min, the reaction mixture was heated to 95 °C and reaction was continued for 24 h. The preparation of the nanocomposite required energy associated with cooling (2 °C, ice and water bath), mechanical stirring: stage I (10 min) and stage II (24 h), heating (95 °C) and centrifugation as well as ethanol for washing and redispersion.

3.4. *Life Cycle Assessment Methodology and Assumptions*

The LCA is a tool for evaluating the environmental impacts of products or processes, analyzing the inputs and the outputs of a defined system. The LCA methodology comprises four steps: scope definition, inventory analysis, impact assessment and interpretation of results. In the first step, the boundaries of the system and the functional unit in each case study are defined. By applying this definition and limits, it is possible to identify mass and energy balances. The next step is inventory analysis, in which information on both input and output flows is compiled and sorted. In the impact assessment stage, the inventory flows are associated with their contributions to the impact categories defined by the methodology used (e.g., CML, IMPACT2001 or ReCiPe). These categories are variable, depending to the method chosen, e.g., climate change (CC), ozone depletion (OD), freshwater eutrophication (FEU), or terrestrial acidification (TA). The values obtained can be analyzed and discussed at the stage of interpretation of results and are useful for comparing different alternatives or for evaluating the environmental impact of a process or product [20].

To improve the comparison between the results, a normalized value for impact can be calculated, as the weighted sum of the impact categories considering their relative contribution. The factors used in this step can be provided using different methodologies. This standardization makes it possible to obtain a global parameter of the environmental impact, allowing the comparison between categories, establishing relative weights between them and improving the understanding of the data.

The LCA performed was based on a cradle-to-gate approach, from the production of the raw materials to the formulation of the NPs. The weight of NP produced per batch was selected as the functional unit (FU), so that it was easier to compare relative consumption of raw materials, water, chemicals, and energy, which were considered primary data obtained from the average values corresponding to different production batches. The experimental procedures for the synthesis of nanoparticles were used to calculate the main inputs and outputs. The Ecoinvent®database (2016) was used to collect background data.

Data uncertainty modeling is an interesting approach for identifying the possible robustness of the analysis and inventory data. Uncertainty in life cycle analysis is often related to the determination of relevant data, e.g., consumption or emissions figures. All the data related to the production of nanoparticles correspond to real data, where the production protocol is presented in the materials and method section. This is why they are representative and accurate data. Another option is to consider the uncertainty related to the secondary data. Since "true" values (especially for background data) are often unknown, it is virtually impossible to completely avoid uncertainty in LCA. Stochastic modeling, which can be performed by Monte Carlo simulation, is a promising technique for addressing data inaccuracy in LCI, but it is beyond the scope of this work.

The characterization factors reported by the ReCiPe methodology for Life Cycle Impact Assessment (LCIA) were used [32], a method based on the ISO standards for LCA [33]. The impact potentials evaluated were: climate change (CC), ozone depletion (OD), terrestrial acidification (TA), freshwater eutrophication (FEU), marine eutrophication (MEU), human toxicity (HT), photochemical oxidants formation (POF), terrestrial ecotoxicity (TE), freshwater ecotoxicity (FE), marine ecotoxicity (ME), water depletion (WD), and fossil depletion (FD). The software SimaPro 8.2.0 (PRé Sustainability, Amersfoort, UT, The Netherlands, 2016) was used for the computational implementation of the inventories.

3.5. Removal of Target Dyes by Fenton and Photo-Fenton Nanocatalysts

Degradation tests were carried out on nanocatalysts with better environmental profiles more environmentally friendly for the degradation of Reactive Blue 19 (RB19), Congo Red (CR), Orange II (OII) and Methyl Green (MG), selected as models of organic pollutants. The oxidative conditions were as follows: 10 mg L^{-1} dye, 300 mg L^{-1} catalyst, 300 mg L^{-1} H_2O_2, in 10 mL H_2O at pH 3 in Fenton applications, or pH 7 in UVA-assisted photocatalysis using a PenRay®Lamp (Analytik Jena, Jena, TH, Germany). For the calibration of the spectrophotometer PowerWave XS2 (BioTek, Winnoski, VT, USA), standard solutions of dyes with different concentrations were prepared and absorbance at the optimal wavelength was correlated with concentration for each dye.

4. Conclusions

An LCA perspective for the production of magnetite-based NPs applied in catalytic oxidation processes for wastewater treatment highlights the importance of assessing the environmental profile associated with the different production routes. Energy and chemical consumption are the dominant impacts during NP synthesis. As for Fenton, the simplest option of sterically stabilized magnetite shows fewer impacts, but the use of the nanocomposites, despite their higher contributions to environmental categories, arises as a much more advisable alternative due to the combination of transformation potential and reactivity. Similarly, in photocatalysis, when it comes to selecting the best option, the nanocomposite has better performance and even faster reaction kinetics compared to Fenton. Specifically, Fe_3O_4@PAA/SBA15 and Fe_3O_4/ZnO magnetic nanocomposites are the most reliable options over the other catalytic supports for Fenton and photo-Fenton processes.

Author Contributions: Investigation and formal analysis, S.F.; data curation and resources, J.G.-R., L.F. and C.V.-V.; writing—original draft preparation and conceptualization, M.T.M. and G.F. All authors have read and agreed to the published version of the manuscript.

Funding: This research was supported by two projects granted by Spanish Ministry of Science and Innovation: MODENA Project CTQ2016-79461-R and CLUSTERCAT Project MAT2015-67458-P, and Fundación Ramón Areces, Spain (Project CIVP18A3940). The authors belong to the Galician Competitive Research Groups ED431C-2017/22 and ED431C-2017/29 and CRETUS Institute.

Conflicts of Interest: The authors declare no conflict of interest.

References

1. Esplugas, S.; Giménez, J.; Contreras, S.; Pascual, E.; Rodríguez, M. Comparison of different advanced oxidation processes for phenol degradation. *Water Res.* **2002**, *36*, 1034–1042. [CrossRef]
2. Bautista, P.; Mohedano, A.F.; Casas, J.A.; Zazo, J.A.; Rodriguez, J.J. An overview of the application of Fenton oxidation to industrial wastewaters treatment. *J. Chem. Technol. Biotechnol.* **2008**, *83*, 1323–1338. [CrossRef]
3. Sayed, M.; Khan, J.A.; Shah, L.A.; Shah, N.S.; Khan, H.M.; Rehman, F.; Khan, A.R.; Khan, A.M. Degradation of quinolone antibiotic, norfloxacin, in aqueous solution using gamma-ray irradiation. *Environ. Sci. Pollut. Res.* **2016**, *23*, 13155–13168. [CrossRef] [PubMed]
4. Yu, K.; Gu, C.; Boyd, S.A.; Liu, C.; Sun, C.; Teppen, B.J.; Li, H. Rapid and Extensive Debromination of Decabromodiphenyl Ether by Smectite Clay-Templated Subnanoscale Zero-Valent Iron. *Environ. Sci. Technol.* **2012**, *46*, 8969–8975. [CrossRef]
5. Peller, J.R.; Mezyk, S.P.; McKay, G.; Watson, E. Hydroxyl Radical Probes for the Comparison of Secondary Treated Wastewaters. *Water Reclam. Sustain.* **2014**, 247–263. [CrossRef]
6. Babuponnusami, A.; Muthukumar, K. A review on Fenton and improvements to the Fenton process for wastewater treatment. *J. Environ. Chem. Eng.* **2014**, *2*, 557–572. [CrossRef]
7. Wang, N.; Zheng, T.; Zhang, G.; Wang, P. A review on Fenton-like processes for organic wastewater treatment. *J. Environ. Chem. Eng.* **2016**, *4*, 762–787. [CrossRef]
8. Lee, K.M.; Lai, C.W.; Ngai, K.S.; Juan, J.C. Recent developments of zinc oxide based photocatalyst in water treatment technology: A review. *Water Res.* **2016**, *88*, 428–448. [CrossRef] [PubMed]
9. Turan, N.B.; Erkan, H.S.; Engin, G.O.; Bilgili, M.S. Nanoparticles in the aquatic environment: Usage, properties, transformation and toxicity—A review. *Process Saf. Environ. Prot.* **2019**, *130*, 238–249. [CrossRef]
10. Brar, S.K.; Verma, M.; Tyagi, R.D.; Surampalli, R.Y. Engineered nanoparticles in wastewater and wastewater sludge—Evidence and impacts. *Waste Manag.* **2010**, *30*, 504–520. [CrossRef]
11. Maity, D.; Choo, S.-G.; Yi, J.; Ding, J.; Xue, J.M. Synthesis of magnetite nanoparticles via a solvent-free thermal decomposition route. *J. Magn. Magn. Mater.* **2009**, *321*, 1256–1259. [CrossRef]
12. Hou, Y.; Yu, J.; Gao, S. Solvothermal reduction synthesis and characterization of superparamagnetic magnetite nanoparticles. *J. Mater. Chem.* **2003**, *13*, 1983–1987. [CrossRef]
13. Kostyukhin, E.M.; Kustov, L.M. Microwave-assisted synthesis of magnetite nanoparticles possessing superior magnetic properties. *Mendeleev Commun.* **2018**, *28*, 559–561. [CrossRef]
14. Monárrez-Cordero, B.E.; Amézaga-Madrid, P.; Hernández-Salcedo, P.G.; Antúnez-Flores, W.; Leyva-Porras, C.; Miki-Yoshida, M. Theoretical and experimental analysis of the aerosol assisted CVD synthesis of magnetite hollow nanoparticles. *J. Alloys Compd.* **2014**, *615*, S328–S334. [CrossRef]
15. Morjan, I.; Alexandrescu, R.; Scarisoreanu, M.; Fleaca, C.; Dumitrache, F.; Soare, I.; Popovici, E.; Gavrila, L.; Vasile, E.; Ciupina, V.; et al. Controlled manufacturing of nanoparticles by the laser pyrolysis: Application to cementite iron carbide. *Appl. Surf. Sci.* **2009**, *255*, 9638–9642. [CrossRef]
16. Akbarzadeh, A.; Samiei, M.; Davaran, S. Magnetic nanoparticles: Preparation, physical properties, and applications in biomedicine. *Nanoscale Res. Lett.* **2012**, *7*, 144. [CrossRef]
17. Parsai, T.; Kumar, A. Understanding effect of solution chemistry on heteroaggregation of zinc oxide and copper oxide nanoparticles. *Chemosphere* **2019**, *235*, 457–469. [CrossRef]
18. Xu, J.-K.; Zhang, F.-F.; Sun, J.-J.; Sheng, J.; Wang, F.; Sun, M. Bio and Nanomaterials Based on Fe3O4. *Molecules* **2014**, *19*, 21506–21528. [CrossRef]
19. Fernández, L.; Gamallo, M.; González-Gómez, M.A.; Vázquez-Vázquez, C.; Rivas, J.; Pintado, M.; Moreira, M.T. Insight into antibiotics removal: Exploring the photocatalytic performance of a Fe3O4/ZnO nanocomposite in a novel magnetic sequential batch reactor. *J. Environ. Manag.* **2019**, *237*, 595–608. [CrossRef]
20. Curran, M.A. Life Cycle Assessment: A review of the methodology and its application to sustainability. *Curr. Opin. Chem. Eng.* **2013**, *2*, 273–277. [CrossRef]

21. International Organization for Standardization. *ISO 14040-Environmental Management—Life Cycle Assessment—Principles and Framework*; International Organization for Standardization: Geneva, Switzerland, 2006.
22. Hischier, R.; Walser, T. Life cycle assessment of engineered nanomaterials: State of the art and strategies to overcome existing gaps. *Sci. Total Environ.* **2012**, *425*, 271–282. [CrossRef] [PubMed]
23. Gavankar, S.; Suh, S.; Keller, A.F. Life cycle assessment at nanoscale: Review and recommendations. *Int. J. Life Cycle Assess.* **2012**, *17*, 295–303. [CrossRef]
24. Upadhyayula, V.K.K.; Meyer, D.E.; Curran, M.A.; Gonzalez, M.A. Life cycle assessment as a tool to enhance the environmental performance of carbon nanotube products: A review. *J. Clean. Prod.* **2012**, *26*, 37–47. [CrossRef]
25. Feijoo, S.; González-García, S.; Moldes-Diz, Y.; Vazquez-Vazquez, C.; Feijoo, G.; Moreira, M.T. Comparative life cycle assessment of different synthesis routes of magnetic nanoparticles. *J. Clean. Prod.* **2017**, *143*, 528–538. [CrossRef]
26. Colilla, M.; Balas, F.; Manzano, M.; Vallet-Regí, M. Novel method to enlarge the surface area of SBA-15. *Chem. Mater.* **2007**, *19*, 3099–3101. [CrossRef]
27. Wang, H.; Chen, Q.W.; Yu, Y.F.; Cheng, K.; Sun, Y.B. Size-and solvent-dependent magnetically responsive optical diffraction of carbon-encapsulated superparamagnetic colloidal photonic crystals. *J. Phys. Chem. C* **2011**, *115*, 11427–11434. [CrossRef]
28. Osorio, Z.V.; Pineiro, Y.; Vazquez, C.; Abreu, C.R.; Perez, M.A.A.; Quintela, M.A.L.; Rivas, J. Magnetic nanocomposites based on mesoporous silica for biomedical applications. *Int. J. Nanotechnol.* **2016**, *13*, 648. [CrossRef]
29. Vargas-Osorio, Z.; González-Gómez, M.A.; Piñeiro, Y.; Vázquez-Vázquez, C.; Rodríguez-Abreu, C.; López-Quintela, M.A.; Rivas, J. Novel synthetic routes of large-pore magnetic mesoporous nanocomposites (SBA-15/Fe3O4) as potential multifunctional theranostic nanodevices. *J. Mater. Chem. B* **2017**, *5*, 9395–9404. [CrossRef]
30. Feldmann, C.; Jungk, H.-O. Polyol-Mediated Preparation of Nanoscale Oxide Particles. *Angew. Chem. Int. Ed.* **2001**, *40*, 359–362. [CrossRef]
31. Chemseddine, A.; Moritz, T. Nanostructuring Titania: Control over Nanocrystal Structure, Size, Shape, and Organization. *Eur. J. Inorg. Chem.* **1999**, *1999*, 235–245. [CrossRef]
32. Goedkoop, M.; Heijungs, R.; Huijbregts, M.; Schryver, A.d.; Strujis, J.; van Zelm, R. *A Life Cycle Impact Assessment Method Which Comprises Harmonised Category Indicators at the Midpoint and the Endpoint Level*; Centrum Milieukunde: Leiden, The Netherlands, 2013.
33. ISO. *ISO 14044*; ISO: Geneva, Switzerland, 2006; ISBN 5935522004.

Article

Biochar-Supported FeS/Fe_3O_4 Composite for Catalyzed Fenton-Type Degradation of Ciprofloxacin

Yue Wang, Xiaoxiao Zhu, Dongqing Feng, Anthony K. Hodge, Liujiang Hu, Jinhong Lü and Jianfa Li *

Department of Chemistry, Shaoxing University, Shaoxing 312000, China; wangyue835@163.com (Y.W.); mymondy@163.com (X.Z.); fengdongqing@yeah.net (D.F.); hodgekanthony2016@gmail.com (A.K.H.); huliujiang@163.com (L.H.); lujhong@aliyun.com (J.L.)

* Correspondence: ljf@usx.edu.cn

Received: 29 October 2019; Accepted: 11 December 2019; Published: 13 December 2019

Abstract: The Fenton-type oxidation catalyzed by iron minerals is a cost-efficient and environment-friendly technology for the degradation of organic pollutants in water, but their catalytic activity needs to be enhanced. In this work, a novel biochar-supported composite containing both iron sulfide and iron oxide was prepared, and used for catalytic degradation of the antibiotic ciprofloxacin through Fenton-type reactions. Dispersion of FeS/Fe_3O_4 nanoparticles was observed with scanning electron microscopy-energy dispersive X-ray spectroscopy (SEM-EDS) and transmission electron microscopy (TEM). Formation of ferrous sulfide (FeS) and magnetite (Fe_3O_4) in the composite was validated by X-ray diffraction (XRD) and X-ray photoelectron spectroscopy (XPS). Ciprofloxacin (initial concentration = 20 mg/L) was completely degraded within 45 min in the system catalyzed by this biochar-supported magnetic composite at a dosage of 1.0 g/L. Hydroxyl radicals (·OH) were proved to be the major reactive species contributing to the degradation reaction. The biochar increased the production of ·OH, but decreased the consumption of H_2O_2, and helped transform Fe^{3+} into Fe^{2+}, according to the comparison studies using the unsupported FeS/Fe_3O_4 as the catalyst. All the three biochars prepared by pyrolysis at different temperatures (400, 500 and 600 °C) were capable for enhancing the reactivity of the iron compound catalyst.

Keywords: biochar; iron mineral; heterogeneous catalysts; antibiotic; advanced oxidation processes; water treatment; organic pollutants; Fenton reactions

---

1. Introduction

Biochar is a kind of carbonaceous material obtained by pyrolysis of biomass feedstock, such as wood processing residues and agricultural wastes. Based on the abundant feedstock sources and flexible preparation processes, the biochar with diverse functions can be developed and used in various processes for the removal of pollutants from water [1,2]. For example, the porous biochar with high specific surface area is a good adsorbent of organic pollutants [3,4], while the biochar rich in alkalis is more suitable for the precipitation of heavy metals (e.g., Pb^{2+} and Cd^{2+}) [5,6]. Due to its low cost, high surface area and good stability, biochar is a promising supporting material comparable with other carbons [7,8], and has been applied to enhance the performance of iron or iron minerals in environmental remediation. First, the biochar can disperse the zero-valent iron nanoparticles and prevent their aggregation, so that the removal efficiency of heavy metals (e.g., Cr(VI) and As(V)) was significantly improved [9,10]. Second, the biochar-supported nanoscale zero-valent iron has shown to be an efficient catalyst for the Fenton-type oxidation of organic pollutants, such as trichloroethylene [11,12], bisphenol A [13], sulfamethazine [14] and ciprofloxacin [15].

Biochar can activate the decomposition of hydrogen peroxide (H_2O_2) or persulfates to produce reactive oxygen species (ROSs) such as hydroxyl radicals (·OH) or sulfate radical anions ($SO_4{\cdot}^-$), which in turn can be used for oxidation of pollutants [16–19]. So, biochar is different from those inactive supporting materials, such as clays, that do not react with H_2O_2. Third, the magnetic biochar containing magnetite (Fe_3O_4) is the most popular biochar-based composite of iron compounds. In comparison to ordinary biochar, the magnetic biochar is easier to be separated and recycled, and showed greater ability to remove pollutants [20,21]. However, most of the previous studies about magnetic biochar focused on the adsorptive removal of pollutants [22]. Its application as the catalyst in Fenton-type systems has attracted the attention of researchers in the most recent years [23–25]. The previous research results indicate that the biochar-supported magnetite nanoparticles are promising catalysts for the degradation of refractory pollutants such as polycyclic aromatic hydrocarbons [24,25] and antibiotics [26–28].

In comparison to iron oxides, iron sulfides (FeS or FeS_2) are often more reactive when used as the catalysts for Fenton-type oxidation of organic pollutants, as both Fe(II) and S(-II)/S(-I) are electron donors for the activation of H_2O_2 [29–31]. FeS is the main component of Mackinawite, a common iron mineral that exists extensively in soil. FeS is also easily synthesized at mild conditions. Recent studies found that FeS is effective for activation of both H_2O_2 and persulfate, for the Fenton-type oxidation of 2,4-dichlorophenoxyacetic acid [32] and p-chloroaniline [33]. However, FeS particles prepared using ordinary methods tend to aggregate, which diminishes the active sites available for surface reactions [34]. Dispersion of FeS particles on biochar has been reported to increase its reactivity for more efficient immobilization of Cr(VI) [35,36]. However, till now there have hardly been any reports in literature about the performance of biochar-based FeS composites for the Fenton-type degradation of pollutants.

In this work, a novel biochar-supported composite containing nanosized FeS and Fe_3O_4 was prepared, and used for catalytic degradation of an organic pollutant through Fenton-type reactions. Ciprofloxacin, a common antibiotic that is frequently found in polluted water, was used as the pollutant model [37–39]. FeS was used here for catalyzing Fenton-type reactions, while Fe_3O_4 for giving the composite magnetism besides catalyzing the reactions. To the best of our knowledge, the combination of both iron sulfide and iron oxide with biochar has never been reported previously. The biochar-supported composite was characterized by scanning electron microscopy-energy dispersive X-ray spectroscopy (SEM-EDS), transmission electron microscopy (TEM), X-ray diffraction (XRD) and X-ray photoelectron spectroscopy (XPS). Its performance on the removal of ciprofloxacin was evaluated by batch degradation experiments, and compared with the unsupported FeS/Fe_3O_4 composite. The reactive oxygen species (ROS) in the Fenton-type systems were investigated, and the influence of biochar and reaction conditions was discussed.

2. Results and Discussion

2.1. Characterizations of the Biochar-Supported Composite

The SEM image in Figure 1a shows the smooth surface of biochar pieces, and that in Figure 1b indicates the rough surface of unsupported FeS/Fe_3O_4 particles with size ranged from 2–30 μm. The loading of iron compound on the biochar did not change the shape of biochar particles (Figure 1c), which have a size of <150 μm in overall. Dispersion of FeS/Fe_3O_4 particles on the biochar can be found in the biochar-supported composite (FeS/Fe_3O_4@BC500), according to that shown in Figure 1c,d, as most of the iron compound particles in the supported composite have a size of <10 μm. The TEM image in Figure 1e further proves the dispersion of nanosized FeS/Fe_3O_4 particles on biochar, while these nanoparticles would aggregate together without the support of biochar as that shown in Figure 1f. The SEM-EDS analysis results (Figure S1) demonstrate that the wood-derived biochar is composed only by C and O, as well as H, according to the elemental analysis results (C 85.3%, O 8.61% and H 2.39% by mass for BC500). No other elements were observed in this biochar, which makes it different from those biochars rich in inorganic minerals [40,41].

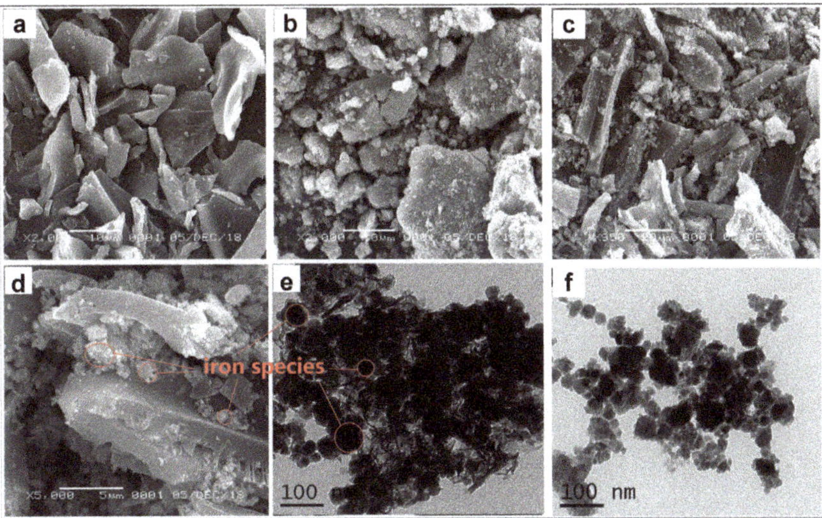

Figure 1. Scanning electron microscopy (SEM) images of (**a**) BC500 biochar (×2000), (**b**) FeS/Fe_3O_4 sample (×2000), (**c**) FeS/Fe_3O_4@BC500 composite (×950), and (**d**) FeS/Fe_3O_4@BC500 composite (×5000). TEM images of (**e**) FeS/Fe_3O_4@BC500 composite, and (**f**) FeS/Fe_3O_4 sample.

Further, no element other than Fe, O and S was observed on the unsupported FeS/Fe_3O_4 sample (Figure S1), indicating its high purity. The EDS analysis result indicates that the S content on the FeS/Fe_3O_4 surface (2.59% by mass) is much lower than that calculated (18.2% by mass) from the theoretical composition (FeS:Fe_3O_4 = 1:1 by mass). This difference should be resulted from the specific preparation procedure of FeS/Fe_3O_4 composite in this work. FeS was synthesized first and formed the inner core of the composite, while Fe_3O_4 deposited itself on the outer surface of these composite particles, accompanying with partial exposure of FeS. Therefore, the higher O content and lower S content were measured with the EDS technique that is a semi-quantitative tool to measure surface composition. The similar EDS result, namely the S content (2.52% by mass) lower than the theoretical value, was obtained on the biochar-supported composite (FeS/Fe_3O_4@BC500) (Figure S1). Here, the theoretical S content should be 6.06% by mass according to the composition of the composite (FeS:Fe_3O_4:BC500 = 1:1:4 by mass). The outer surface composing of iron oxide (Fe_3O_4) is beneficial for the stability of our catalyst in open air, as Fe_3O_4 is more resistant to oxidation by air than FeS.

The uniform distribution of S, O and Fe elements according to the EDS mapping (Figure 2) implies the formation of both iron sulfide and iron oxide that are further identified by XRD analysis (Figure 3a). The peaks belonging to both Fe_3O_4 and FeS were observed on FeS/Fe_3O_4 and FeS/Fe_3O_4@BC500 samples. The fewer and weaker peaks of FeS are related to the fact that most of the FeS is located in the inner core of particles, as discussed above about the SEM-EDS results. XPS analysis (Figure 3b) further proves the formation of two iron compounds. As that can be seen, the Fe 2p peaks at 710.1 and 711.4 eV are ascribed to the binding energies of Fe(II) of FeS and Fe_3O_4, respectively, and the S 2p peaks at 161.9 and 163.5 eV correspond to the binding energy of S^{2-} and $S_n{}^{2-}$. The peaks belonging to S^0 and S–O should be resulted from the partial oxidation of S^{2-} after exposure to air. Such oxidation may also lead to the weak signals of the S element in EDS analysis (Figure 2) and the weak FeS signal in XRD analysis (Figure 3a). The formation of Fe_3O_4 endows the magnetism, as the FeS/Fe_3O_4@BC500 powder was attracted by a magnet according that shown in Figure 3c.

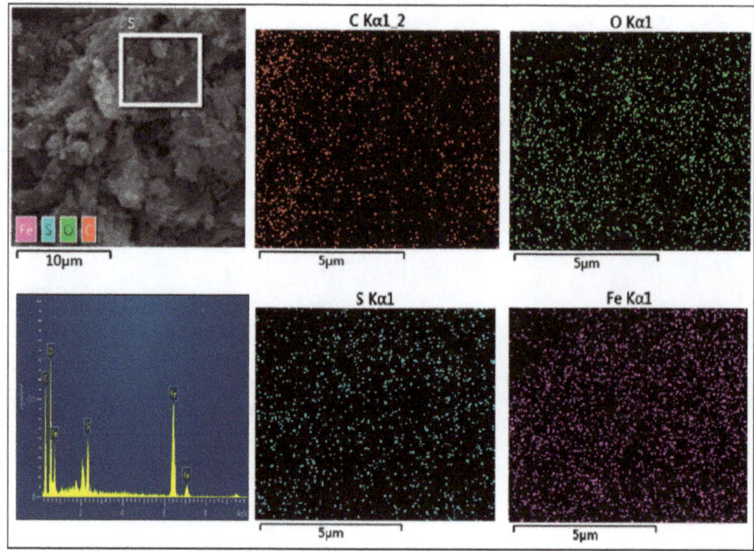

Figure 2. Electron microscopy-energy dispersive (EDS) mapping of C, O, S and Fe elements on the surface of the FeS/Fe_3O_4@BC500 composite.

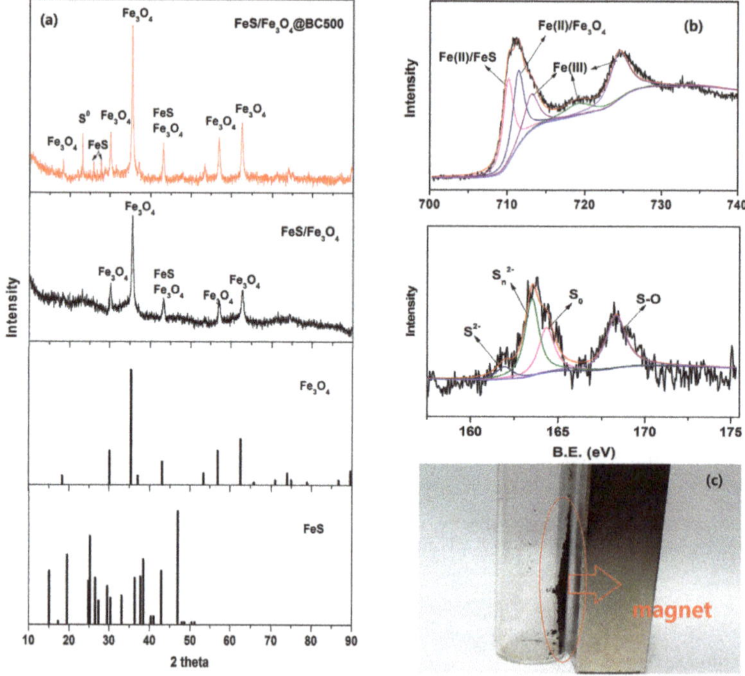

Figure 3. (**a**) X-ray diffractometry (XRD) patterns of the FeS/Fe_3O_4 and FeS/Fe_3O_4@BC500 samples; (**b**) XPS spectra of Fe 2p and S 2p regions of the FeS/Fe_3O_4@BC500 composite; and (**c**) the photo of FeS/Fe_3O_4@BC500 composite attracted by magnet.

2.2. Comparison of Various Catalysis Systems for Removal of Ciprofloxacin

The removal of ciprofloxacin by Fenton-type systems using various catalysts, including FeS/Fe_3O_4@BC500 composite, BC500 biochar and the FeS/Fe_3O_4 sample, is shown in Figure 4a. Here the dosages of iron compound (FeS/Fe_3O_4) and biochar (BC500) were set at 0.33 g/L and 0.67 g/L, respectively, equivalent to the theoretical contents in 1.0 g/L of FeS/Fe_3O_4@BC500 (1:1:4). The results of control experiments without the addition of H_2O_2 are also included in the same figure, and at least 20% of ciprofloxacin was removed by the catalysts alone in the initial stage (<2 min). The removal of ciprofloxacin by BC500 alone should be attributed to adsorption, as the biochar has a high specific surface area of 298.8 m^2/g. However, there was more ciprofloxacin removed by the two iron-containing samples (FeS/Fe_3O_4 and FeS/Fe_3O_4@BC500) than that by the biochar alone, despite that the specific surface areas of FeS/Fe_3O_4 (32.5 m^2/g) and FeS/Fe_3O_4@BC500 (120.6 m^2/g) are much smaller than the BC500 biochar. The reason should be related to the degradation of ciprofloxacin in these two systems, besides surface adsorption. Both iron sulfide and magnetite have been reported to be capable for activating dissolved O_2 in water, which results in the production of hydroxyl radicals (·OH) that are highly reactive for oxidative degradation of organic compounds [42–44]. For validating such a guess, the removal of ciprofloxacin by FeS/Fe_3O_4 or FeS/Fe_3O_4@BC500 alone was conducted under nitrogen atmosphere. Fewer pollutants were removed under nitrogen atmosphere than in the open air (Figure S2a), confirming the contribution of dissolved O_2 to the degradation of ciprofloxacin. Further, there was still a significant part (~20%) of ciprofloxacin removed by FeS/Fe_3O_4 or FeS/Fe_3O_4@BC500 in the absence of dissolved O_2 (Figure S2a), which should be attributed to adsorption. Rakshit et al. [45] has reported that ciprofloxacin was strongly adsorbed by nanomagnetite, and the quantity of ciprofloxacin adsorbed was as high as 0.04 mmol/g. The adsorption isotherms of ciprofloxacin by FeS/Fe_3O_4 and FeS/Fe_3O_4@BC500 (Figure S2b) also proved the role of these two materials as adsorbent of ciprofloxacin.

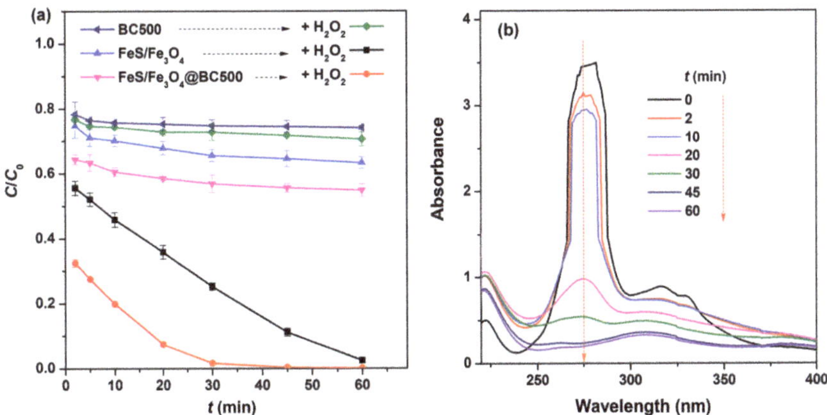

Figure 4. (**a**) Removal of ciprofloxacin (C_0 = 0.06 mmol/L) in the Fenton-type systems catalyzed by BC500 biochar, FeS/Fe_3O_4 sample and FeS/Fe_3O_4@BC500 composite, as well as the control experiments without H_2O_2; (**b**) Change of ultraviolet absorbance spectra of solution samples taken at different reaction time from the system catalyzed by FeS/Fe_3O_4@BC500 composite. Dosage of catalysts: FeS/Fe_3O_4@BC500 1.0 g/L, FeS/Fe_3O_4 0.33 g/L, or BC500 0.67 g/L; and dosage of H_2O_2 = 4.0 mmol/L. pH_0 = 3.0.

Anyway, the much quicker removal of ciprofloxacin was observed in the two H_2O_2-containing systems catalyzed by the FeS/Fe_3O_4 and FeS/Fe_3O_4@BC500 composites. In particular, nearly complete degradation of ciprofloxacin could be achieved within 45 min in the FeS/Fe_3O_4@BC500 system. The steep drop of ciprofloxacin concentration in the initial stage (<2 min) should be due to both the adsorption and degradation of ciprofloxacin. The higher removal of ciprofloxacin in the FeS/Fe_3O_4@BC500 system than that in the FeS/Fe_3O_4 system is partially resulted from the well dispersion of FeS/Fe_3O_4 particles on biochar, because there are more reactive sites on FeS/Fe_3O_4@BC500 available for the degradation of pollutant. Further, the aqueous samples taken at various reaction times were analyzed with ultraviolet absorbance, so as to validate that the removal of ciprofloxacin was resulted from a degradation reaction. The spectra shown in Figure 4b indicate the gradual attenuation of peak absorption at ~275 nm with the reaction proceeding, accompanying with a emerging peak at 225 nm. This peak at ~275 nm is characteristic of quinoline structure in ciprofloxacin molecules. Its attenuation and disappearance after a reaction of 30 min indicates the breakdown of quinoline structure. No other strong absorbance peaks, except the emerging one at 225 nm, were observed in the solution samples after this reaction of 30 min, implying almost complete transformation of ciprofloxacin into intermediates (Figure S3) such as short-chain acids and CO_2, according to the previous research reports [37,38,46].

Repetitive experiments using the recycled composite catalysts were performed to assess their reusability (Figure 5). The dropped removal efficiency was observed in the FeS/Fe_3O_4 system since the second run, indicating the reduced catalytic activity. In comparison, the FeS/Fe_3O_4@BC500 composite maintained its catalytic activity in the second run, and the gradually decreasing removal of ciprofloxacin was found in this system after the third run. So, due to the dispersion of FeS/Fe_3O_4 on biochar, the supported FeS/Fe_3O_4@BC500 catalyst exhibited higher reusability and activity in the repetitive experiments than FeS/Fe_3O_4. The reduced activity of both catalysts should be resulted from the leaching of iron specie, because Fe content on the surface of the recycled catalysts decreased in comparison with the pristine catalysts according to SEM-EDS analysis (Figure S4 vs. Figures S1b,c). The reduced activity should also be resulted from the contamination of catalyst by reaction intermediates formed on the surface [30,32,33]. Anyway, no significant change on the mineral compositions was observed on the catalysts after reaction, according to the XRD analysis results (Figure S5).

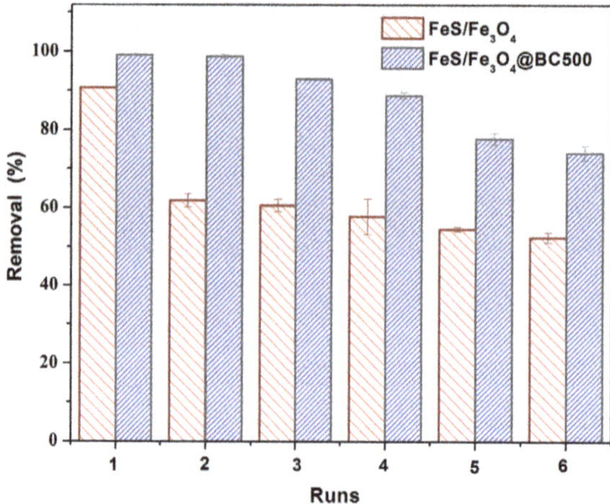

Figure 5. Removal of ciprofloxacin (C_0 = 0.06 mmol/L) in repetitive experiments after each run in the Fenton-type systems using different catalysts. Dosage: FeS/Fe_3O_4@BC500 = 1.0 g/L, FeS/Fe_3O_4 = 0.33 g/L and H_2O_2 = 4.0 mmol/L. pH_0 = 3.0.

2.3. Investigation about the Reactive Species

Methanol is a common ·OH scavenger that is frequently used for determining the major reactive oxygen species in the Fenton or Fenton-type systems. In this study, the degradation reaction of ciprofloxacin was dramatically suppressed when methanol (40 mmol/L) was added in the systems catalyzed by FeS/Fe_3O_4 and FeS/Fe_3O_4@BC500, according to the results shown in Figure 6a. The results confirm that ·OH is the reactive species that has made the major contribution for the oxidation of ciprofloxacin, which is consistent with previous reports about the Fenton-type reactions catalyzed by iron minerals [27,29,31,32,39,47,48]. For further investigating the role of biochar, the cumulative production of ·OH in the two Fenton-type systems was measured using benzoic acid (BA) as the probe. The results in Figure 6b indicate that there were more ·OH produced in the FeS/Fe_3O_4@BC500 system than in the FeS/Fe_3O_4 system. The cumulative amount of ·OH produced after 60 min in the former system is 0.657 mmol/L, which is 45% higher than that in the later system. In contrast, much fewer consumption of H_2O_2 was observed in the FeS/Fe_3O_4@BC500 system than that in the FeS/Fe_3O_4 system (Figure 6b). Based on the molar yield of ·OH, the utilization efficiency of H_2O_2 was enhanced from 17% in the FeS/Fe_3O_4 system to 83% in the FeS/Fe_3O_4@BC500 system with the aid of biochar. The dramatically enhanced yield of ·OH should be related to the stabilization of radicals by resonance effect from aromatic carbon and/or hydroquinone/quinone structure of biochar [49], which contributed to the accumulation of persistent free radicals in biochar according to previous studies [16,17].

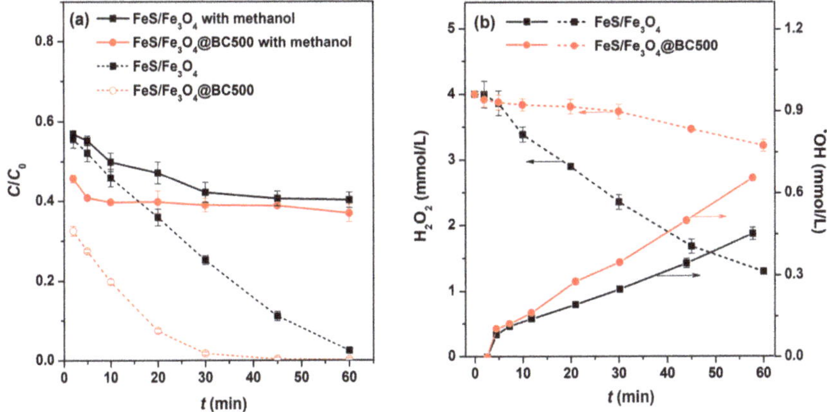

Figure 6. (**a**) Influence of methanol additive (40 mmol/L) on the removal of ciprofloxacin (C_0 = 0.06 mmol/L) in the two Fenton-type systems, and the dash lines represent those without additive; (**b**) Production of ·OH and consumption of H_2O_2 in the two systems. Dosage of catalysts: FeS/Fe_3O_4@BC500 1.0 g/L and FeS/Fe_3O_4 0.33 g/L, dosage of H_2O_2 = 4.0 mmol/L, and pH_0 = 3.0.

Furthermore, in the solution of the FeS/Fe_3O_4 system, Fe^{2+} content increased first, but then decreased in the later stage (after reaction of 30 min) (Figure 7a), which should be resulted from the oxidation of Fe^{2+}, as the total Fe in solution increased in the later stage. In contrast, both Fe^{2+} and total Fe in the solution of the FeS/Fe_3O_4@BC500 system maintained at a relatively constant level (Figure 7a). Further, the released total Fe in the FeS/Fe_3O_4@BC500 system is much fewer than that in the FeS/Fe_3O_4 system. After a reaction of 60 min, the total Fe in the FeS/Fe_3O_4@BC500 system is 2.1 mg/L, only 43% of that released in the FeS/Fe_3O_4 system (Figure 7a).

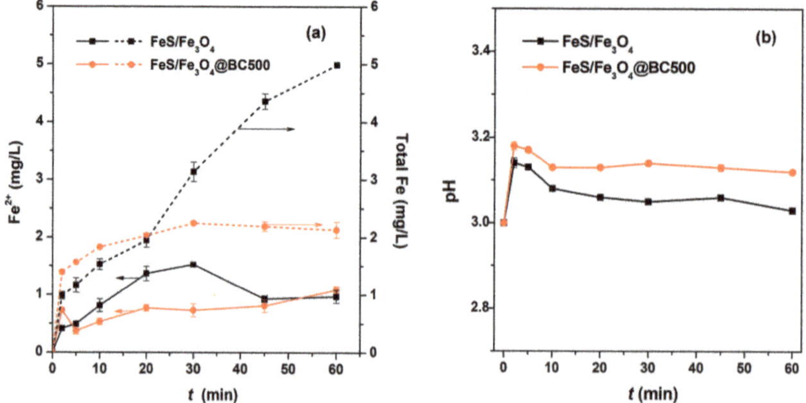

Figure 7. (**a**) Release of total Fe and Fe^{2+}, and (**b**) pH change during the reaction in the two Fenton-type systems (C_0 = 0.06 mmol/L). Dosage of catalysts: FeS/Fe_3O_4@BC500 1.0 g/L and FeS/Fe_3O_4 0.33 g/L, dosage of H_2O_2 = 4.0 mmol/L, and pH_0 = 3.0.

The results indicate the biochar-supported composite is more stable than the unsupported one, which is beneficial for reusability of the biochar-supported catalyst. The reasons should be mainly attributed to the function of biochar as an electron donor that can help transform Fe^{3+} into Fe^{2+} [50]. There are many hydroquinone (Ar–OH) and quinone groups (Ar–C=O) on the biochar according to its infrared spectra shown in Figure S6. The conversion of biochar hydroquinone and quinone groups would donate electrons that promoted the Fe^{2+}/Fe^{3+} cycle [50,51], which is beneficial for increasing the yield of ·OH at the less consumption of H_2O_2 (Figure 6b). In addition, pH change during the reaction in the two Fenton-type systems was also recorded, and results in Figure 7b indicate that pH in the FeS/Fe_3O_4@BC500 system is somewhat higher than that in the FeS/Fe_3O_4 system. The results indicate that the accelerated removal of ciprofloxacin in the former system should not attribute to the pH change, as it is well known that pH close to 3.0 is beneficial for quicker Fenton-type reactions.

2.4. The Biochar's Influence to the Removal of Ciprofloxacin

Two other biochar samples prepared at different pyrolysis temperatures (400 and 600 °C) were used instead of BC500 for preparation of the biochar-supported composites. As that can be seen in Figure 8a, all the composites have demonstrated to be more reactive than FeS/Fe_3O_4 for removal of ciprofloxacin. Among the three biochar-based composites, FeS/Fe_3O_4@BC400 shows to be less reactive than the others (Figure 8a). This should be related to ·OH production in the corresponding Fenton-type systems, because fewer ·OH radicals were produced in the system catalyzed by FeS/Fe_3O_4@BC400 (Figure 8b). The reason should be attributed to the properties of BC400 biochar different from the other two biochars. Previous studies have shown that the biochar prepared at lower temperatures (<400 °C) has more un-carbonized organic residues such as polycyclic aromatic hydrocarbons [52]. These organic residues might compete with the target pollutant (ciprofloxacin) for the consumption of reactive species [53]. The BC600 biochar alone can remove more ciprofloxacin, according to the results in Figure 8a. Such a higher removal should be resulted from the more adsorption of ciprofloxacin by this biochar that has higher specific surface area (378 m^2/g). The relatively higher reactivity of FeS/Fe_3O_4@BC500 and FeS/Fe_3O_4@BC600 composites in catalyzing the degradation of ciprofloxacin (Figure 8a) should be related to the higher aromaticity of biochar prepared at higher pyrolysis temperature. Because more aromatic structure in biochar favors the resonance stabilization of free radicals [40,54].

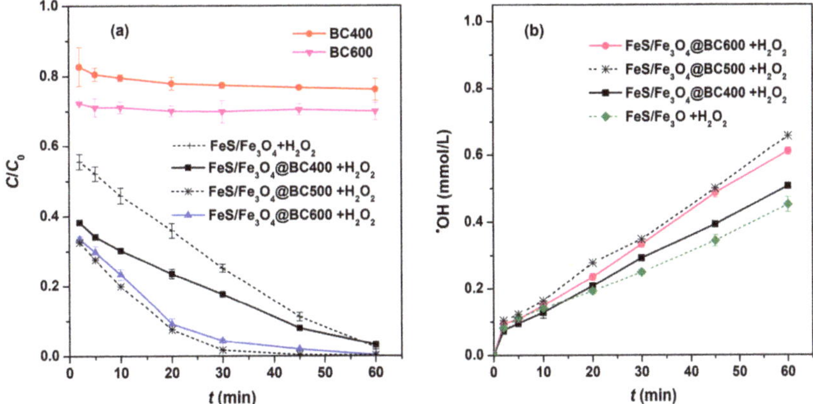

Figure 8. (**a**) Removal of ciprofloxacin (C_0 = 0.06 mmol/L) in the Fenton-type systems catalyzed by composite based on different biochar (BC400, BC500 and BC600), and the control experiments using biochar alone; (**b**) Production of ·OH in the systems using composite catalysts based on different biochar. Dosage: the composite catalyst = 1.0 g/L, the biochar = 0.67 g/L, H_2O_2 = 4.0 mmol/L. pH_0 = 3.0.

In general, the biochars prepared at temperatures other than 500 °C were also applicable as support of iron compounds for enhancing their catalytic activity, while the reasons for the different performance of biochar prepared at different temperatures deserve further investigations.

2.5. *Influence of Reaction Conditions*

The delayed removal of ciprofloxacin was observed with increasing pH. The much quicker removal of ciprofloxacin was obtained at pH_0 = 2.0 or 3.0 (Figure 9a), indicating that release of Fe^{2+} was the dominant factor for this Fenton-type system. As the increase of pH suppressed the dissolution of Fe^{2+} from the solid catalyst, the Fenton reactions and degradation of ciprofloxacin were slowed down. Furthermore, the optimum molar H_2O_2/ciprofloxacin ratio for removal of ciprofloxacin was in the range of 33.3 to 66.7 (and the dosage of H_2O_2 was 2 to 4 mmol/L), according to the results shown in Figure 9b. This range is 0.71 to 1.42 times the stoichiometric ratio for total mineralization of ciprofloxacin by Fenton's oxidation, which is 47 according to a previous study [46]. Fewer H_2O_2 would limit the amount of ·OH produced, while excess H_2O_2 dosage is unfavorable (Figure 9b), because the excess H_2O_2 would react with ·OH (Equation (1)). Such reaction will produce less oxidizing species (HO_2·) together with extra consumption of ·OH, so that the oxidation of ciprofloxacin by ·OH was impacted adversely.

$$H_2O_2 + \cdot OH \rightarrow HO_2\cdot + H_2O \quad (1)$$

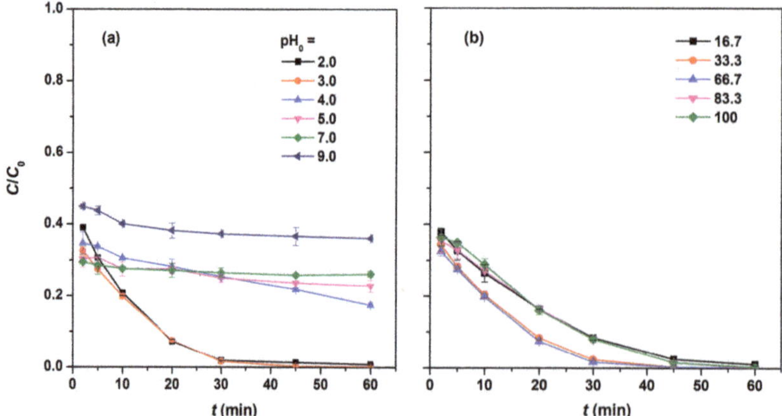

Figure 9. Removal of ciprofloxacin (C_0 = 0.06 mmol/L) in the Fenton-type system catalyzed by FeS/Fe_3O_4@BC500 (1.0 g/L) at: (**a**) different pH_0 at the same molar H_2O_2/ciprofloxacin ratio of 66.7, and (**b**) different molar H_2O_2/ciprofloxacin ratio at the same pH_0 of 3.0.

On the basis of above results and discussion about the reactive species and influencing factors, the Fenton-type reactions catalyzed by the biochar-supported FeS/Fe_3O_4 composite can be outlined. First, dissolution of Fe^{2+} from the catalyst (Equation (2)) in acidic pH was the primary step to initiate the following reaction for ·OH production (Equation (3)), so the performance of this catalyst in neutral and alkalescent pH was impacted adversely. Second, conversion of biochar hydroquinone (BCHQ) to quinone (BCQ) facilitated the regeneration of Fe^{2+} (Equation (4)), which replaced the reaction (5) essential for Fe^{2+} regeneration in the traditional Fenton system. Therefore, the extra consumption of H_2O_2 by side reactions, such as that which occurred in Equation (5), was reduced in the system catalyzed by the FeS/Fe_3O_4@BC500 composite (Figure 6b).

$$\mathrm{BC} \equiv \mathrm{Fe(II)} \rightarrow \mathrm{Fe}^{2+}(\mathrm{dissolution}) \tag{2}$$

$$\mathrm{Fe}^{2+} + \mathrm{H_2O_2} \rightarrow \mathrm{Fe}^{3+} + \mathrm{OH}^- + \cdot\mathrm{OH} \tag{3}$$

$$\mathrm{Fe}^{3+} + \mathrm{BCHQ} \rightarrow \mathrm{Fe}^{2+} + \mathrm{BCQ} \tag{4}$$

$$\mathrm{Fe}^{3+} + \mathrm{H_2O_2} \rightarrow \mathrm{Fe}^{2+} + \mathrm{HO_2}\cdot + \mathrm{H}^+ \tag{5}$$

3. Materials and Methods

3.1. Materials

Biochar was prepared by slow pyrolysis of pinewood (*Pinus radiata*) sawdust in a tube furnace (Φ 8 cm × 100 cm in length) according to the procedure reported previously [4,5,55]. The sawdust was washed with pure water, and air-dried to constant weight. The quartz vessel loaded with clean sawdust was put into the furnace filled with a continuous nitrogen flow of 200 mL/min. Then the furnace was heated to a set temperature (400, 500 or 600 °C) at a heating rate of 8 °C/min, followed by a holding time of 3.0 h. The solid residue in the furnace was taken out from the tube after the furnace was cooled to room temperature, and soaked in pure water for 24 h to remove any soluble matter from the biochar. The biochar produced at a different pyrolysis temperature is labeled as BC400, BC500 or BC600, respectively, where the suffix indicates the pyrolysis temperature in degrees Celsius. The biochar particles passing through a 100-mesh sieve (particle size <150 μm) were used in the following experiments.

Ciprofloxacin ($C_{17}H_{18}FN_3O_3$) of 98% purity was purchased from Aladdin Co. Ltd. (Shanghai, China). Ferrous sulfate heptahydrate ($FeSO_4{\cdot}7H_2O$), ferric sulfate ($Fe_2(SO_4)_3$), sodium sulfide nonahydrate ($Na_2S{\cdot}9H_2O$) and sodium hydroxide (NaOH) were all of analytical grade and purchased from Sinopharm Chemical Reagent Co., Ltd. (Shanghai, China). The 18.2 MΩ·cm pure water was obtained from the Milli-Q Gradient (Millipore, Billerica, MA, USA) pure water system and used for preparing solution of ciprofloxacin and other solutions.

3.2. Preparation of the Biochar-Supported FeS/Fe_3O_4 Composites

The biochar-supported FeS/Fe_3O_4 composite was prepared by three steps. First, the biochar (e.g., BC500) was dispersed in 150 mL of deoxygened pure water, and 0.1 mol/L Na_2S was dissolved in the dispersion, followed by dropwise addition of 34.5 mL of 0.6 mol/L $FeSO_4$ solution; thereby FeS was precipitated. Second, 38 mL of 0.15 mol/L $Fe_2(SO_4)_3$ solution was added dropwise, after then 45.5 mL of 1.0 mol/L NaOH solution was added under stirring, so that iron oxide precipitated. Third, the mixture was sealed and aged in a warm water (70 °C) bath for 48 h. The solid product was filtered out and dried in a vacuum. The as-prepared product has a mass FeS/Fe_3O_4 ratio of 1:1, which was chosen as a compromise for achieving both high catalysis reactivity and magnetism. According to our preliminary experiments, the higher FeS ratio is favorable for high reactivity, but also leads to poor magnetism. The biochar dosage used for preparing the composite follows the mass FeS/Fe_3O_4/biochar ratio of 1:1:4. Lower biochar dosage is unfavorable for the dispersion of iron species, while higher biochar dosage results in declined reactivity of the composite catalyst. The biochar-supported composites were labeled as FeS/Fe_3O_4@BC400, FeS/Fe_3O_4@BC500 and FeS/Fe_3O_4@BC600, according to the biochar used for preparing the composite. For a comparison study, the unsupported composite of mass FeS/Fe_3O_4 ratio = 1:1 was prepared by same steps without the addition of biochar.

3.3. Characterizations of the Biochar and Composites

The specific surface area (SA) of the samples (biochar and composites) was calculated using the Brunauer–Emmett–Teller (BET) method based on the N_2 adsorption/desorption isotherms that were measured at 77K in a Tristar II 3020 surface area and porosity analyzer (Micromeritics, Norcross, GA, USA), after vacuum degassing at 473 K for 6 h. Scanning electron microscopy-electron microscopy-energy dispersive (SEM-EDS) analysis was operated on a JSM-6360LV scanning electron microscope (JEOL, Tokyo, Japan), equipped with an X-act energy dispersive X-ray spectrometer (Oxford, Abingdon, UK). TEM was observed on a JEM-1011 transmission electron microscope (JEOL, Tokyo, Japan). An X-ray diffractometry (XRD) study was performed on a D/MAX 3A (Rigaku, Tokyo, Japan) equipment with CuKα radiation and a goniometer rate of 4 °/min. XPS was recorded on a Thermo ESCALAB 250 system (Waltham, MA, USA) with a monochromatized Al Kα X-ray source ($h\nu$ = 1486.6 eV) operated at a power of 150 W. The binding energies of photoelectrons were corrected by C 1s peak at 284.8 eV.

3.4. Degradation Experiments

Degradation experiments were conducted in the dark in a 100 mL conical flask containing 50 mL of ciprofloxacin solution and the catalyst. The flask was put in a thermostatic oscillator (PSE-T150A, China) at 25 ± 1 °C. H_2O_2 was quickly added into the stirring suspension to initiate the reaction. The initial concentration (C_0) of ciprofloxacin was 20 mg/L (0.06 mmol/L), and the typical dosage of H_2O_2 was 4 mmol/L, unless specified otherwise. The initial pH (pH_0) was adjusted with dilute H_2SO_4 solution to be 3.0 ± 0.1. At the predetermined reaction time (t, min), 0.3 mL of reaction solution was sampled, and put into methanol of the same volume to quench the reaction, then the solution was filtered by 0.22 μm membrane for subsequent analysis. All the experiments were carried out in triplicates.

Different catalysts, including the biochar-supported composite (e.g., FeS/Fe_3O_4@BC500) of 1.0 g/L, the FeS/Fe_3O_4 sample of 0.33 g/L (equivalent to the amount of iron species used in the supported composite), and the biochar of 0.67 g/L were used for comparison studies. The various dosages of H_2O_2 (1.0, 2.0, 4.0, 5.0 and 6.0 mmol/L, equivalent to the molar H_2O_2/ciprofloxacin ratio of 16.7, 33.3, 66.7, 83.3 and 100, respectively), and different pH_0 (2.0, 3.0, 4.0, 5.0, 7.0 and 9.0 adjusted with 0.1 mol/L H_2SO_4 or NaOH solutions) were used for investigating the influence of reaction conditions. Reusability of the catalysts was tested by separation of the solid out from the reaction system first, and the recycled catalysts were repetitively used by addition of ciprofloxacin solution and H_2O_2.

3.5. Analytical Methods

Ciprofloxacin concentration in aqueous samples was analyzed by using a HPLC system (Shimadzu LC-20A, Kyoto, Japan) equipped with a multi-wavelength UV detector. The separation was performed on an ODS-2 column (5 µm, 150 × 4.6 mm) with a flow rate of 1.0 mL /min at 25 °C. The mobile phase was a mixture of acetonitrile and 25 mmol/L H_3PO_4 solution (V/V = 16/84), and the detection wavelength was set at 275 nm (the peak absorbance of ciprofloxacin). The concentration of aqueous Fe^{2+} was measured at 510 nm using the 1,10-phenanthroline method on a UV-vis spectrophotometer (Spectrum 1920, Shanghai, China). The total aqueous Fe was examined by atomic absorption spectrometry (Shimadzu AA-7000, Japan). The accumulative production of ·OH was measured using benzoic acid as the probe, and the production of 1 mol of p-hydroxybenzoic acid (*p*-HBA) from this reaction was reported to consume 5.87 mol of ·OH [56]. The *p*-HBA was measured with HPLC using the mixture of acetonitrile/water (containing 0.15% acetic acid) = 60/40 (*v/v*) as the mobile phase at a flow rate of 1 mL/min, and using UV detection at 254 nm. H_2O_2 was analyzed by the Titanate method at 551 nm using the UV-vis spectrophotometer.

4. Conclusions

The biochar-supported composite containing both iron sulfide (FeS) and iron oxide (Fe_3O_4) ($FeS:Fe_3O_4$ = 1:1 by mass) was prepared, and has shown to be an efficient catalyst for Fenton-type oxidation of the antibiotic ciprofloxacin. The biochar made the iron species well dispersed, and enhanced its reactivity for catalyzing the degradation of ciprofloxacin. Ciprofloxacin (C_0 = 0.06 mmol/L) could be completely degraded within 45 min in the Fenton-type system catalyzed by FeS/Fe_3O_4@BC500 (1.0 g/L) with the molar H_2O_2/ciprofloxacin ratio of 33.3 to 66.7. The biochar increased the production of ·OH by 45% with a much less consumption of H_2O_2. The biochar also promoted the Fe^{2+}/Fe^{3+} cycle by acting as electron donor, which is suggested to be attributed to the hydroquinone/quinone structure in biochar. However, the reactivity of this composite catalyst depended mostly on the dissolution of Fe, which impacted its performance in neutral and alkalescent pH.

Supplementary Materials: The following are available online at http://www.mdpi.com/2073-4344/9/12/1062/s1, Figure S1: SEM-EDS results of BC500 biochar, FeS/Fe_3O_4 sample and FeS/Fe_3O_4@BC500 composite, Figure S2: Removal and adsorption isotherms of ciprofloxacin by FeS/Fe_3O_4 and FeS/Fe_3O_4@BC500 composite under nitrogen atmosphere, Figure S3: High performance liquid chromatogram of solution samples taken after different reaction time from the FeS/Fe_3O_4@BC500 system, Figure S4: SEM-EDS results of FeS/Fe_3O_4 and FeS/Fe_3O_4@BC500 composite after reaction, Figure S5: XRD patterns of FeS/Fe_3O_4 and FeS/Fe_3O_4@BC500 samples after reaction, Figure S6: Infrared spectra of BC500 biochar.

Author Contributions: Conceptualization, J.L. (Jianfa Li) and L.H.; Methodology, Y.W. and X.Z.; Investigation, Y.W., X.Z., D.F., A.K.H. and J.L. (Jinhong Lü); Validation, D.F.; Writing—Original Draft Preparation, Y.W.; Writing—Review and Editing, J.L. (Jianfa Li); Supervision, J.L. (Jianfa Li); Funding Acquisition, J.L. (Jianfa Li).

Funding: This research was funded by the National Natural Science Foundation of China, grant number 21777103.

Conflicts of Interest: The authors declare no conflict of interest.

References

1. Mukherjee, A.; Zimmerman, A.R.; Harris, W. Surface chemistry variations among a series of laboratory-produced biochars. *Geoderma* **2011**, *163*, 247–255. [CrossRef]
2. Chen, Z.; Xiao, X.; Chen, B.; Zhu, L. Quantification of chemical states, dissociation constants and contents of oxygen-containing groups on the surface of biochars produced at different temperatures. *Environ. Sci. Technol.* **2014**, *49*, 309–317. [CrossRef] [PubMed]
3. Ahmad, M.; Lee, S.S.; Rajapaksha, A.U.; Vithanage, M.; Zhang, M.; Cho, J.S.; Lee, S.-E.; Ok, Y.S. Trichloroethylene adsorption by pine needle biochars produced at various pyrolysis temperatures. *Bioresour. Technol.* **2013**, *143*, 615–622. [CrossRef] [PubMed]
4. Zhu, X.; Li, C.; Li, J.; Xie, B.; Lü, J.; Li, Y. Thermal treatment of biochar in the air/nitrogen atmosphere for developed mesoporosity and enhanced adsorption to tetracycline. *Bioresour. Technol.* **2018**, *263*, 475–482. [CrossRef] [PubMed]
5. Xu, X.; Cao, X.; Zhao, L.; Wang, H.; Yu, H.; Gao, B. Removal of Cu, Zn, and Cd from aqueous solutions by the dairy manure-derived biochar. *Environ. Sci. Pollut. Res.* **2013**, *20*, 358–368. [CrossRef] [PubMed]
6. Zhang, T.; Zhu, X.; Shi, L.; Li, J.; Li, S.; Lü, J.; Li, Y. Efficient removal of lead from solution by celery-derived biochars rich in alkaline minerals. *Bioresour. Technol.* **2017**, *235*, 185–192. [CrossRef]
7. Outsiou, A.; Frontistis, Z.; Ribeiro, R.S.; Antonopoulou, M.; Konstantinou, I.K.; Silva, A.M.; Faria, J.L.; Gomes, H.T.; Mantzavinos, D. Activation of sodium persulfate by magnetic carbon xerogels (CX/CoFe) for the oxidation of bisphenol A: Process variables effects, matrix effects and reaction pathways. *Water Res.* **2017**, *124*, 97–107. [CrossRef]
8. Soares, O.S.G.P.; Rodrigues, C.S.; Madeira, L.M.; Pereira, M.F.R. Heterogeneous Fenton-Like degradation of *p*-nitrophenol over tailored carbon-based materials. *Catalysts* **2019**, *9*, 258. [CrossRef]
9. Zhou, Y.; Gao, B.; Zimmerman, A.R.; Chen, H.; Zhang, M.; Cao, X. Biochar-supported zerovalent iron for removal of various contaminants from aqueous solutions. *Bioresour. Technol.* **2014**, *152*, 538–542. [CrossRef]
10. Diao, Z.H.; Du, J.J.; Jiang, D.; Kong, L.J.; Huo, W.Y.; Liu, C.M.; Wu, Q.H.; Xu, X.R. Insights into the simultaneous removal of Cr^{6+} and Pb^{2+} by a novel sewage sludge-derived biochar immobilized nanoscale zero valent iron: Coexistence effect and mechanism. *Sci. Total Environ.* **2018**, *642*, 505–515. [CrossRef]
11. Yan, J.; Han, L.; Gao, W.; Xue, S.; Chen, M. Biochar supported nanoscale zerovalent iron composite used as persulfate activator for removing trichloroethylene. *Bioresour. Technol.* **2015**, *175*, 269–274. [CrossRef] [PubMed]
12. Li, Z.; Sun, Y.; Yang, Y.; Han, Y.; Wang, T.; Chen, J.; Tsang, D.C. Biochar-supported nanoscale zero-valent iron as an efficient catalyst for organic degradation in groundwater. *J. Hazard. Mater.* **2020**, *383*, 121240. [CrossRef] [PubMed]
13. Liu, C.M.; Diao, Z.H.; Huo, W.Y.; Kong, L.J.; Du, J.J. Simultaneous removal of Cu^{2+} and bisphenol A by a novel biochar-supported zero valent iron from aqueous solution: Synthesis, reactivity and mechanism. *Environ. Pollut.* **2018**, *239*, 698–705. [CrossRef] [PubMed]
14. Deng, J.; Dong, H.; Zhang, C.; Jiang, Z.; Cheng, Y.; Hou, K.; Zhang, L.; Fan, C. Nanoscale zero-valent iron/biochar composite as an activator for Fenton-like removal of sulfamethazine. *Sep. Purif. Technol.* **2018**, *202*, 130–137. [CrossRef]
15. Mao, Q.; Zhou, Y.; Yang, Y.; Zhang, J.; Liang, L.; Wang, H.; Luo, S.; Luo, L.; Jeyakumar, P.; Ok, Y.S.; et al. Experimental and theoretical aspects of biochar-supported nanoscale zero-valent iron activating H_2O_2 for ciprofloxacin removal from aqueous solution. *J. Hazard. Mater.* **2019**, *380*, 120848. [CrossRef]
16. Fang, G.; Gao, J.; Liu, C.; Dionysiou, D.D.; Wang, Y.; Zhou, D. Key role of persistent free radicals in hydrogen peroxide activation by biochar: Implications to organic contaminant degradation. *Environ. Sci. Technol.* **2014**, *48*, 1902–1910. [CrossRef]
17. Fang, G.; Liu, C.; Gao, J.; Dionysiou, D.D.; Zhou, D. Manipulation of persistent free radicals in biochar to activate persulfate for contaminant degradation. *Environ. Sci. Technol.* **2015**, *49*, 5645–5653. [CrossRef]
18. Kemmou, L.; Frontistis, Z.; Vakros, J.; Manariotis, I.D.; Mantzavinos, D. Degradation of antibiotic sulfamethoxazole by biochar-activated persulfate: Factors affecting the activation and degradation processes. *Catal. Today* **2018**, *313*, 128–133. [CrossRef]
19. Magioglou, E.; Frontistis, Z.; Vakros, J.; Manariotis, I.D.; Mantzavinos, D. Activation of persulfate by biochars from valorized olive stones for the degradation of sulfamethoxazole. *Catalysts* **2019**, *9*, 419. [CrossRef]

20. Chen, B.; Chen, Z.; Lv, S. A novel magnetic biochar efficiently sorbs organic pollutants and phosphate. *Bioresour. Technol.* **2011**, *102*, 716–723. [CrossRef]
21. Wang, S.; Gao, B.; Zimmerman, A.R.; Li, Y.; Ma, L.; Harris, W.G.; Migliaccio, K.W. Removal of arsenic by magnetic biochar prepared from pinewood and natural hematite. *Bioresour. Technol.* **2015**, *175*, 391–395. [CrossRef] [PubMed]
22. Thines, K.R.; Abdullah, E.C.; Mubarak, N.M.; Ruthiraan, M. Synthesis of magnetic biochar from agricultural waste biomass to enhancing route for waste water and polymer application: A review. *Renew. Sustain. Energy Rev.* **2017**, *67*, 257–276. [CrossRef]
23. Ouyang, D.; Yan, J.; Qian, L.; Chen, Y.; Han, L.; Su, A.; Zhang, W.; Ni, H.; Chen, M. Degradation of 1, 4-dioxane by biochar supported nano magnetite particles activating persulfate. *Chemosphere* **2017**, *184*, 609–617. [CrossRef] [PubMed]
24. Dong, C.D.; Chen, C.W.; Hung, C.M. Synthesis of magnetic biochar from bamboo biomass to activate persulfate for the removal of polycyclic aromatic hydrocarbons in marine sediments. *Bioresour. Technol.* **2017**, *245*, 188–195. [CrossRef]
25. Dong, C.D.; Chen, C.W.; Kao, C.M.; Chien, C.C.; Hung, C.M. Wood-biochar-supported magnetite nanoparticles for remediation of PAH-contaminated estuary sediment. *Catalysts* **2018**, *8*, 73. [CrossRef]
26. Yi, Y.; Tu, G.; Zhao, D.; Tsang, P.E.; Fang, Z. Pyrolysis of different biomass pre-impregnated with steel pickling waste liquor to prepare magnetic biochars and their use for the degradation of metronidazole. *Bioresour. Technol.* **2019**, *289*, 121613. [CrossRef]
27. Yi, Y.; Tu, G.; Tsang, P.E.; Fang, Z. Insight into the influence of pyrolysis temperature on Fenton-like catalytic performance of magnetic biochar. *Chem. Eng. J.* **2020**, *380*, 122518. [CrossRef]
28. Nguyen, V.T.; Hung, C.M.; Nguyen, T.B.; Chang, J.H.; Wang, T.H.; Wu, C.H.; Lin, Y.L.; Chen, C.W.; Dong, C.D. Efficient heterogeneous activation of persulfate by iron-modified biochar for removal of antibiotic from aqueous solution: A case study of tetracycline removal. *Catalysts* **2019**, *9*, 49. [CrossRef]
29. Liu, W.; Wang, Y.; Ai, Z.; Zhang, L. Hydrothermal synthesis of FeS_2 as a high-efficiency Fenton reagent to degrade alachlor via superoxide-mediated Fe(II)/Fe(III) cycle. *ACS Appl. Mater. Interfaces* **2015**, *7*, 28534–28544. [CrossRef]
30. Diao, Z.H.; Liu, J.J.; Hu, Y.X.; Kong, L.J.; Jiang, D.; Xu, X.R. Comparative study of Rhodamine B degradation by the systems pyrite/H_2O_2 and pyrite/persulfate: Reactivity, stability, products and mechanism. *Sep. Purif. Technol.* **2017**, *184*, 374–383. [CrossRef]
31. Zhao, L.; Chen, Y.; Liu, Y.; Luo, C.; Wu, D. Enhanced degradation of chloramphenicol at alkaline conditions by S(-II) assisted heterogeneous Fenton-like reactions using pyrite. *Chemosphere* **2017**, *188*, 557–566. [CrossRef] [PubMed]
32. Chen, H.; Zhang, Z.; Yang, Z.; Yang, Q.; Li, B.; Bai, Z. Heterogeneous Fenton-like catalytic degradation of 2,4-dichlorophenoxyacetic acid in water with FeS. *Chem. Eng. J.* **2015**, *273*, 481–489. [CrossRef]
33. Yuan, Y.; Tao, H.; Fan, J.; Ma, L. Degradation of p-chloroaniline by persulfate activated with ferrous sulfide ore particles. *Chem. Eng. J.* **2015**, *268*, 38–46. [CrossRef]
34. Gong, Y.; Tang, J.; Zhao, D. Application of iron sulfide particles for groundwater and soil remediation: A review. *Water Res.* **2016**, *89*, 309–320. [CrossRef] [PubMed]
35. Lyu, H.; Tang, J.; Huang, Y.; Gai, L.; Zeng, E.Y.; Liber, K.; Gong, Y. Removal of hexavalent chromium from aqueous solutions by a novel biochar supported nanoscale iron sulfide composite. *Chem. Eng. J.* **2017**, *322*, 516–524. [CrossRef]
36. Lyu, H.; Zhao, H.; Tang, J.; Gong, Y.; Huang, Y.; Wu, Q.; Gao, B. Immobilization of hexavalent chromium in contaminated soils using biochar supported nanoscale iron sulfide composite. *Chemosphere* **2018**, *194*, 360–369. [CrossRef]
37. Yahya, M.S.; Oturan, N.; El Kacemi, K.; El Karbane, M.; Aravindakumar, C.T.; Oturan, M.A. Oxidative degradation study on antimicrobial agent ciprofloxacin by electro-Fenton process: Kinetics and oxidation products. *Chemosphere* **2014**, *117*, 447–454. [CrossRef]
38. Diao, Z.H.; Xu, X.R.; Jiang, D.; Li, G.; Liu, J.J.; Kong, L.J.; Zuo, L.Z. Enhanced catalytic degradation of ciprofloxacin with FeS_2/SiO_2 microspheres as heterogeneous Fenton catalyst: Kinetics, reaction pathways and mechanism. *J. Hazard. Mater.* **2017**, *327*, 108–115. [CrossRef]

39. Hassani, A.; Karaca, M.; Karaca, S.; Khataee, A.; Açışlı, Ö.; Yılmaz, B. Preparation of magnetite nanoparticles by high-energy planetary ball mill and its application for ciprofloxacin degradation through heterogeneous Fenton process. *J. Environ. Manag.* **2018**, *211*, 53–62. [CrossRef]
40. Huang, D.; Luo, H.; Zhang, C.; Zeng, G.; Lai, C.; Cheng, M.; Wang, R.; Deng, R.; Xue, W.; Gong, X.; et al. Nonnegligible role of biomass types and its compositions on the formation of persistent free radicals in biochar: Insight into the influences on Fenton-like process. *Chem. Eng. J.* **2019**, *361*, 353–363. [CrossRef]
41. Li, J.; Pan, L.; Yu, G.; Xie, S.; Li, C.; Lai, D.; Li, Z.; You, F.; Wang, Y. The synthesis of heterogeneous Fenton-like catalyst using sewage sludge biochar and its application for ciprofloxacin degradation. *Sci. Total Environ.* **2019**, *654*, 1284–1292. [CrossRef] [PubMed]
42. Cheng, D.; Yuan, S.; Liao, P.; Zhang, P. Oxidizing impact induced by mackinawite (FeS) nanoparticles at oxic conditions due to production of hydroxyl radicals. *Environ. Sci. Technol.* **2016**, *50*, 11646–11653. [CrossRef] [PubMed]
43. Zhang, P.; Yuan, S.; Liao, P. Mechanisms of hydroxyl radical production from abiotic oxidation of pyrite under acidic conditions. *Geochim. Cosmochim. Acta* **2016**, *172*, 444–457. [CrossRef]
44. Ardo, S.G.; Nélieu, S.; Ona-Nguema, G.; Delarue, G.; Brest, J.; Pironin, E.; Morin, G. Oxidative degradation of nalidixic acid by nano-magnetite via Fe^{2+}/O_2-mediated reactions. *Environ. Sci. Technol.* **2015**, *49*, 4506–4514. [CrossRef]
45. Rakshit, S.; Sarkar, D.; Elzinga, E.J.; Punamiya, P.; Datta, R. Mechanisms of ciprofloxacin removal by nano-sized magnetite. *J. Hazard. Mater.* **2013**, *246–247*, 221–226. [CrossRef]
46. Gupta, A.; Garg, A. Degradation of ciprofloxacin using Fenton's oxidation: Effect of operating parameters, identification of oxidized by-products and toxicity assessment. *Chemosphere* **2018**, *193*, 1181–1188. [CrossRef]
47. Liu, X.; Zhang, Q.; Yu, B.; Wu, R.; Mai, J.; Wang, R.; Chen, L.; Yang, S.T. Preparation of $Fe_3O_4/TiO_2/C$ nanocomposites and their application in Fenton-like catalysis for dye decoloration. *Catalysts* **2016**, *6*, 146. [CrossRef]
48. Zhang, Y.; Tran, H.P.; Hussain, I.; Zhong, Y.; Huang, S. Degradation of p-chloroaniline by pyrite in aqueous solutions. *Chem. Eng. J.* **2015**, *279*, 396–401. [CrossRef]
49. Zhu, X.; Li, J.; Xie, B.; Feng, D.; Li, Y. Accelerating effects of biochar for pyrite-catalyzed Fenton-like oxidation of herbicide 2,4-D. *Chem. Eng. J.* **2020**, 123605. [CrossRef]
50. Qin, Y.; Zhang, L.; An, T. Hydrothermal carbon-mediated Fenton-like reaction mechanism in the degradation of alachlor: Direct electron transfer from hydrothermal carbon to Fe (III). *ACS Appl. Mater. Interfaces* **2017**, *9*, 17115–17124. [CrossRef]
51. Chen, R.; Pignatello, J.J. Role of quinone intermediates as electron shuttles in Fenton and photoassisted Fenton oxidations of aromatic compounds. *Environ. Sci. Technol.* **1997**, *31*, 2399–2406. [CrossRef]
52. Fabbri, D.; Rombolà, A.G.; Torri, C.; Spokas, K.A. Determination of polycyclic aromatic hydrocarbons in biochar and biochar amended soil. *J. Anal. Appl. Pyrolysis* **2013**, *103*, 60–67. [CrossRef]
53. Zhao, X.; Qin, L.; Gatheru Waigi, M.; Cheng, P.; Yang, B.; Wang, J.; Ling, W. Removal of bound PAH residues in contaminated soils by Fenton oxidation. *Catalysts* **2019**, *9*, 619. [CrossRef]
54. Yang, J.; Pan, B.; Li, H.; Liao, S.; Zhang, D.; Wu, M.; Xing, B. Degradation of *p*-nitrophenol on biochars: Role of persistent free radicals. *Environ. Sci. Technol.* **2015**, *50*, 694–700. [CrossRef]
55. Li, J.; Li, Y.; Wu, Y.; Zheng, M. A comparison of biochars from lignin, cellulose and wood as the sorbent to an aromatic pollutant. *J. Hazard. Mater.* **2014**, *280*, 450–457. [CrossRef]
56. Tong, M.; Yuan, S.; Ma, S.; Jin, M.; Liu, D.; Cheng, D.; Liu, X.; Gan, Y.; Wang, Y. Production of abundant hydroxyl radicals from oxygenation of subsurface sediments. *Environ. Sci. Technol.* **2016**, *50*, 214–221. [CrossRef]

Article

Wet Peroxide Oxidation of Paracetamol Using Acid Activated and Fe/Co-Pillared Clay Catalysts Prepared from Natural Clays

Adriano Santos Silva [1,2], Marzhan Seitovna Kalmakhanova [3], Bakytgul Kabykenovna Massalimova [3], Juliana G. Sgorlon [4], Jose Luis Diaz de Tuesta [1,2,*] and Helder Teixeira Gomes [1,2,*]

1 Centro de Investigação de Montanha (CIMO), Instituto Politécnico de Bragança, 5300-253 Bragança, Portugal
2 Laboratory of Separation and Reaction Engineering—Laboratory of Catalysis and Materials (LSRE-LCM), Faculdade de Engenharia, Universidade do Porto, 4200-465 Porto, Portugal
3 Department of Chemistry and Chemical Engineering, M.Kh. Dulati Taraz State University, Tole bi 63, 080012 Taraz, Kazakhstan
4 Universidade Tecnológica Federal do Paraná, R. Marcílio Dias, 635-Jardim Paraiso, Apucarana PR 86812-460, Brasil
* Correspondence: jl.diazdetuesta@ipb.pt (J.L.D.d.T.); htgomes@ipb.pt (H.T.G.)

Received: 29 July 2019; Accepted: 19 August 2019; Published: 22 August 2019

Abstract: Many pharmaceuticals have been recently identified at trace levels worldwide in the aquatic environment. Among them, the highly consumed paracetamol (PCM), an analgesic and antipyretic drug, is largely being accumulated in the aquatic environment due to inefficient removal by conventional sewage treatment plants. This work deals with the treatment of PCM, used as a model pharmaceutical contaminant of emerging concern, by catalytic wet peroxide oxidation using clay-based materials as catalysts. The catalysts were prepared from natural clays, extracted from four different deposits using acid-activated treatment, calcination, and pillarization with Fe and Co. Pillared clays show the highest catalytic activity owing to the presence of metals, allowing to remove completely the PCM after 6 h under the following operating conditions: C_{PCM} = 100 mg L^{-1}, $C_{H_2O_2}$ = 472 mg L^{-1}, C_{cat} = 2.5 g L^{-1}, initial pH = 3.5 and T = 80 °C. The prepared materials presented high stability since leached iron was measured at the end of reaction and found to be lower than 0.1 mg L^{-1}.

Keywords: advanced oxidation process; wastewater treatment; contaminants of emerging concern; pharmaceuticals; environmental catalysis

---

1. Introduction

In recent years, and especially after the development of sophisticated analytical techniques, many pharmaceuticals have been identified at trace levels (ng L^{-1}–mg L^{-1}) worldwide in the aquatic environment [1]. Municipal wastewater treatment plants (WWTPs) are considered the main sources of these pollutants as they are not generally prepared to deal with these complex substances, and thus they are usually ineffective in their complete removal [1–4]. Despite the low concentration of drugs contained in those effluents, their continuous input constitutes an important environmental threat, given their persistence and hazardous nature [5–8].

The presence of pharmaceuticals, even in trace concentrations, affects the quality of water and constitutes a risk of toxicity for the ecosystems and living organisms. A number of effects, such as the development of antibiotic-resistant bacteria in the aquatic environment [7], fish reproduction changes due to the presence of estrogenic compounds [5] and specific inhibition of photosynthesis in algae caused by β-blockers have been reported [6]. Moreover, according to most recent works, the biological effects of low-dose complex micropollutant mixtures are still underestimated [8]. This also constitutes a public health problem since pharmaceuticals have even been found in drinking water supplies [9]. Consequently, new regulation for micropollutant discharge and monitoring has recently started in different countries [10,11].

So far, there are no legal requirements for the discharge of these ubiquitous and biologically active substances, but this scenario is expected to change in the next few years. The European Union (EU) has recently approved a watch list of 17 substances, among them 7 pharmaceuticals, for their monitoring in the EU-water basins (Decision 2015/495/EC) [11]. Those substances showing a significant risk will be potentially listed as priority pollutants.

Among pharmaceutical compounds that can cause pollution of water, paracetamol (PCM) deserves particular attention, since it has recently been discovered as a potential pollutant of waters [12–16]. PCM is an analgesic and antipyretic drug that is largely accumulated in the aquatic environment due to its inefficient removal by conventional sewage treatment plants, also representing an important material for the industry of manufacturing of azo dyes and photographic chemicals [17]. The concern about the environmental impact of its biodegradation products has been growing, because of its hepatotoxicity and the possibility of those products to be toxic or hazardous in trace amounts [12]. It is clear that optimization of WWTPs by including efficient tertiary treatments to create an effective barrier to micropollutants emission is a social responsibility and a task of high priority. Successful results have been reported at the laboratory scale for the elimination of a wide range of pharmaceuticals by Fenton-like processes [18]. The Fenton process has been successful tested in the treatment of pharmaceutical compounds [18,19]. However, the Fenton process produces large amounts of ferrous iron sludge, and an additional process is required. The immobilization of iron onto a solid support in the so-called catalytic wet peroxide oxidation (CWPO) has proved to be an interesting strategy to overcome those limitations. Although this technology represent an interesting alternative, it has been scarcely studied so far for micropollutants abatement [18,20]. Clays play a prominent role as a catalyst in the field of organic pollutants removal by CWPO [21–25]. In this regard, natural clays can be chemically modified to increase their performance by different methods, such as activation or pillarization. On the one hand, the acid activation of a clay consists of the release of metal cations from the layered structure of the clay, creating Lewis and Brönsted sites and leading to an increment of the surface area [26,27]. On the other hand, the process of pillarization consists of the intercalation of the cations present in a previously selected pillaring solution into the interlayer space of the clays, which involves the natural substitution of exchangeable cations present between the sheets of clays [28]. Both activation and pillarization methods can lead to an increase in the catalytic activity of the clays, but clay-based materials have not been assessed in the treatment of contaminants of emerging concern, such as pharmaceutical compounds.

This work deals with the CWPO of PCM, used as pharmaceutical model compound, with modified clay-based catalysts prepared from natural clays extracted from the deposits of four different regions of Kazakhstan. PCM removal and mineralization is addressed with natural, acid-activated, calcined and pillared clays.

2. Results and Discussion

2.1. Characterization of Materials

2.1.1. Textural Properties

The nitrogen adsorption isotherms at 77 K obtained for the prepared clay-based materials are depicted in Figure 1. All materials show similar N_2 sorption isotherms, classified as Type II, as typically found for non-macroporous adsorbents according to the current IUPAC classification [29], following the revisions of 1985 IUPAC recommendations on physisorption isotherms. All samples also show a similar hysteresis loop (only shown in Figure 1 for the KO-PILC sample), classified as Type H3 by IUPAC [29].

Loops of this type are attributed to non-rigid aggregates of plate-like particles (e.g., certain clays) but also to materials with pore networks consisting of macropores not completely filled with condensate. As can be observed, the KO-PILC sample is able to adsorb more nitrogen on its surface when compared to the other pillared clays shown in Figure 1a. However, there are no significant differences between the Kokshetau-based clays prepared from KO-N shown in Figure 1b. The acid activated clay (KO-A) is able to adsorb more nitrogen, resulting in the sample with the best textural properties (Table 1), namely the highest BET surface area and total pore volume (28 $m^2 g^{-1}$ and 94.6 $mm^3 g^{-1}$, respectively). The development of porosity by the acid activation of natural clays could be expected to be higher than observed, but natural clays used in this work possess high amount of impurities in the form of quartz (>45%), limiting the porous development [21]. The pillarization of these clays does not result in an increment of the specific surface area or pore volume, also as a consequence of the quartz content of the natural clays employed, as discussed in previous works [21].

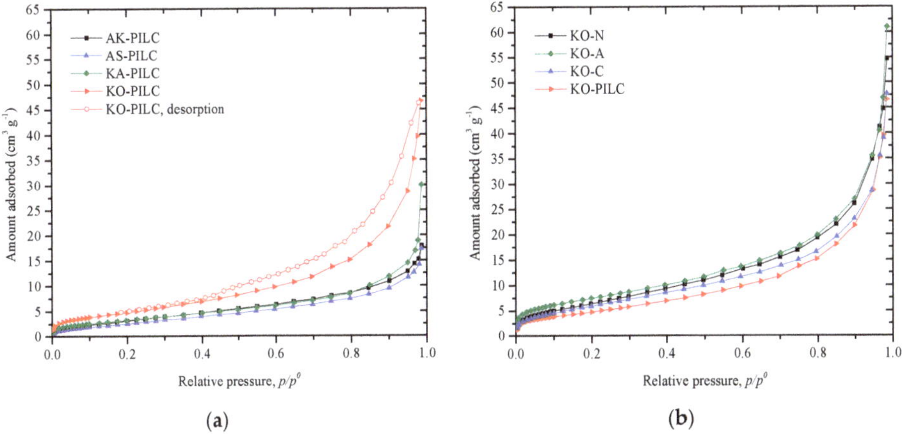

Figure 1. N_2 adsorption isotherms at 77 K of (**a**) pillared clays and (**b**) Kokshetau-based clays.

Table 1. Textural properties of the clay-based samples.

Sample	S_{BET} ($m^2 g^{-1}$)	V_{Total} ($mm^3 g^{-1}$)
KO-N	26	69.2
KO-A	28	94.6
KO-C	24	74.0
KO-PILC	19	72.3
AK-PILC	13	27.8
AS-PILC	11	22.1
KA-PILC	13	46.6

2.1.2. Surface Composition

Figure 2 gathers the XRD diffractograms of the Kokshetau and Karatau based samples. As observed, both clays present the typical reflection of montmorillonite [30]. As the samples were provided from natural deposits, it was expected that its crystal compositions were formed not only by one clay mineral, but by multiple phases. In this sense, it is also possible to find traces of saponite [31], kaolinite [32] and muscovite [33] in the sample KA-N. The material KO-N also evidences the presence of kaolinite [31]. Both natural samples also present a peak at 26.7° related to the presence of impurities of quartz (SiO_2) [34,35]. Even knowing that the natural clays present not only montmorillonite in its crystal composition, the intensities of the peaks in the positions referred to this mineral are higher than the intensities assigned to the other phases, suggesting that the clays are mainly composed by montmorillonite. Therefore, both natural clays can be classified as bentonite, which is the denomination given to the class of clays composed by a mixture of different clay minerals, with a major composition of montmorillonite [36]. It is possible to observe that the signal for SiO_2 and metal oxides such as aluminum oxide and iron (III) oxide decreased in the diffractogram obtained with KO-A when compared to that of KO-N. This can be explained by the fact that the acid treatment washes partially the impurities of SiO_2 from the natural clay, also leaching a small amount of the metals. In the diffractogram of KO-C, the signals attributed to iron (III) and alumina oxides were the same as those obtained in the diffractogram of KO-N, and the signal attributed to SiO_2 had a small decrease. Finally, in the diffractogram of KO-PILC, the signal for iron (III) oxide was significantly higher than in that observed for the natural sample KO-N, confirming the successful incorporation of iron in the clay structure. For the pillared sample KA-PILC it is possible to observe the decrease for the signal of the SiO_2 impurities, and that the signal of the iron (III) oxide increased significantly, putting in evidence the incorporation of iron in the material. In fact, the signal attributed to iron (III) oxide in this sample was higher than in the others.

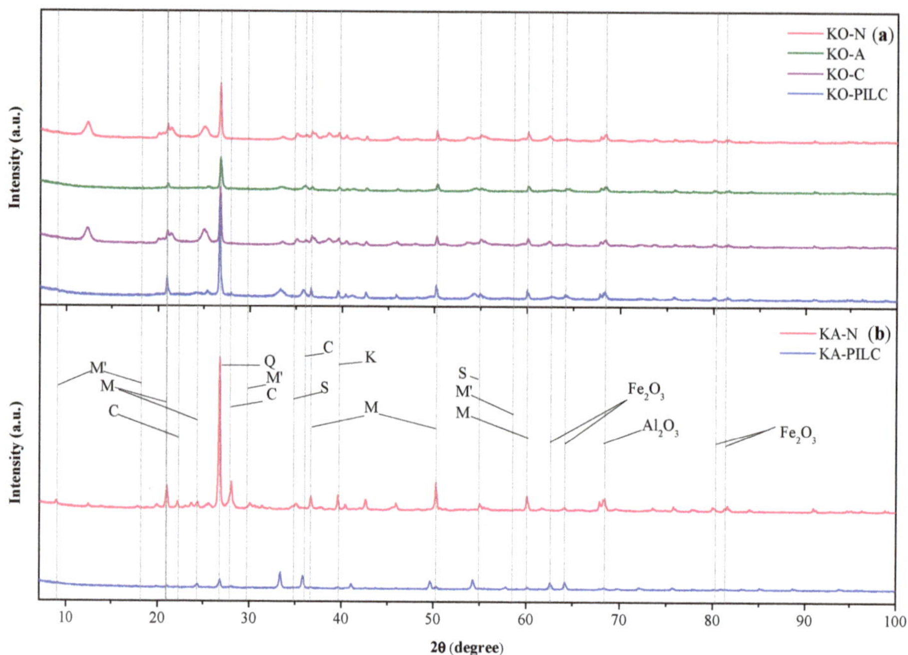

Figure 2. XRD spectra of (**a**) Kokshetau and (**b**) Karatau based clays (M = Montmorillonite, S = Saponite, K = Kaolinite, M′ = Muscovite, C = Calcite, Q = Quartz).

Besides all the differences between the signals in both diffractograms of the Kokshetau and Karatau samples, it is interesting to observe that the signal for montmorillonite, kaolinite, saponite, and muscovite did not change significantly in the different samples. This suggests that the main structure of the clay is stable, although passing through some structural changes [37].

The FT-IR spectra obtained from the analysis of all the natural clays and of the corresponding prepared pillared clay samples are depicted in Figure 3. As can be observed, the natural clays from Akzhar, Asa and Karatau (AK-N, AS-N and KA-N, respectively) show a band close to 1450 cm^{-1}. As detected in a previous work [21], this band is due to the presence of calcite in the natural materials, disappearing after the pillarization process because of the exchange between calcium and the pillaring metals [38]. The band at 870 cm^{-1} appearing in the natural samples can be ascribed to the Al–Mg–OH bending vibrations. The disappearance of this band in the spectra obtained with the pillared samples is due to the fracture of these bonds [39] and can also be ascribed to the exchange of the cation Mg^{2+} by the pillaring procedure. The bands observed in the range of 1000–1025 cm^{-1} is present in the spectra of all the samples, and represent the stretching vibrations of the Si–O bond group [40]. The band observed in the range of wavenumbers from 776 to 780 cm^{-1} is attributed to the presence of the quartz impurity [41]. The band observed close to 530 cm^{-1} is associated with the bending vibrations of the group Si–O–Mg [34]. At 470 cm^{-1} it is also possible to observe another band that is related to the presence of bending vibrations of Si–O–Fe bonds [42,43], which is present in all samples (pillared and natural) as a consequence of the presence of iron in natural clays, as reported previously [21].

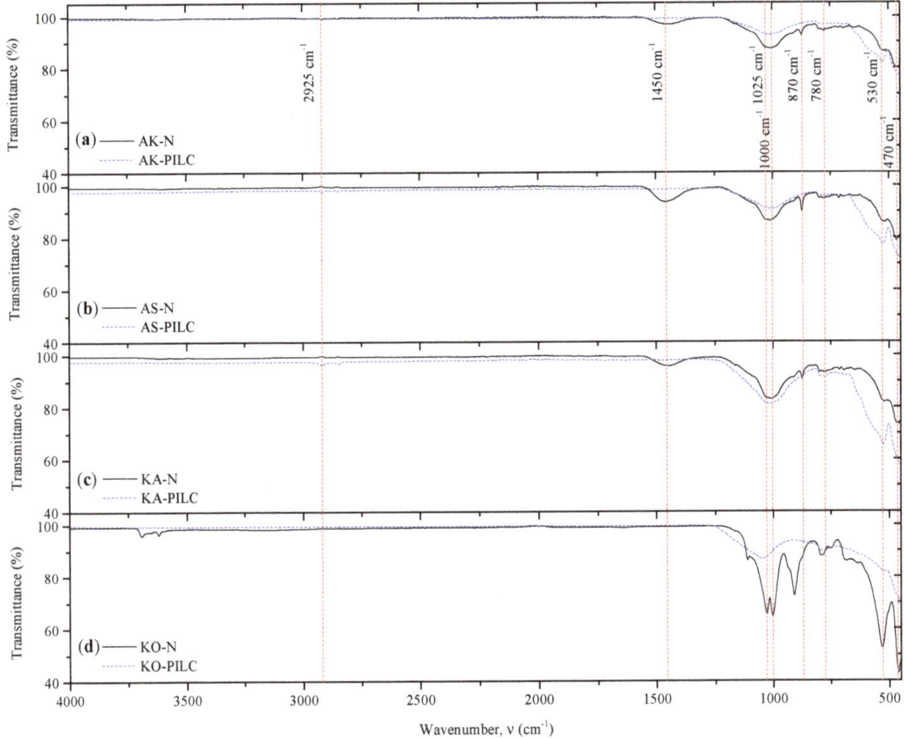

Figure 3. FT-IR spectra of (**a**) Akzhar, (**b**) Asa, (**c**) Karatau and (**b**) Kokshetau based clays in natural and pillared form.

2.1.3. Acid-Base Characterization

The acid–base properties and the pH of the point of zero charge (pH_{PZC}) of the clay-based materials are gathered in Table 2. It can be concluded that all the clay-based materials possess a neutral character around 7–8 of pH_{PZC}.

Table 2. Acid-base characterization of the clay-based materials.

Sample	pH_{PZC}	Acidity (μmol g^{-1})	Basicity (μmol g^{-1})
KO-N	7.8	350	245
KO-A	7.2	987	614
KO-C	8.0	338	270
KO-PILC	7.4	950	652
AK-PILC	7.2	812	538
AS-PILC	7.6	475	372
KA-PILC	7.4	687	627

Among the modified Kokshetau clays, it is possible to observe that the KO-PILC and the KO-A possess more than 950 and 600 μmol g^{-1} respectively of acidic and basic functionalities, whereas the natural and the calcined samples (KO-N and KO-C, respectively) present an acidity and a basicity lower than 350 and 270 μmol g^{-1}, respectively. Thus, the acidity and basicity are increased up to three times after the pillarization or acid activation treatments of the natural clay (KO-N). Both treatments lead to similar acidity (950 and 987 μmol g^{-1} for KO-PILC and KO-A, respectively) and basicity (652 and 614 μmol g^{-1} for KO-PILC and KO-A, respectively), concluding that no significant changes are found in this sense regarding the treatment process.

The pillared clays prepared form the different natural clays also present differences between them. The pillared clay prepared from the Kokshetau natural clay (KO-PILC) presents the highest acidity and basicity (950 and 652 μmol g^{-1}, respectively). The acidity of pillared clays was found to decrease in the following order: KO-PILC (950 μmol g^{-1}) > AK-PILC (812 μmol g^{-1}) > KA-PILC (687 μmol g^{-1}) > AS-PILC (475 μmol g^{-1}), whereas the value of the basicity diminishes as follow: KO-PILC (652 μmol g^{-1}) > KA-PILC (627 μmol g^{-1}) > AK-PILC (538 μmol g^{-1}) > AS-PILC (372 μmol g^{-1}).

The acidity and basicity of a catalyst play an important role in the decomposition of H_2O_2 and in the CWPO of pollutants, as observed in previous works related to the CWPO of phenol compounds with carbon-based catalysts [44]. In works related to the CWPO with clay-based catalysts, a correlation between the acidity of the catalysts and their performance in the oxidation process has also been observed [22–24]. However, the cited previous studies used a qualitative methodology, based on FTIR analysis of base compound-saturated clays, and acidity and basicity were not quantified.

2.2. CWPO of Paracetamol

Figure 4 shows the relative concentration of PCM, H_2O_2 and TOC upon reaction time with the non-pillared clays prepared from the natural clay extracted from the Kokshetau deposit, viz. KO-N, KO-C and KO-A. As observed, all materials are catalytic active in the CWPO of PCM and allow to remove more than 34% of PCM after 24 h, whereas the non-catalytic run lead to a conversion of 20% after 24 h of reaction time at same operating conditions. The removal of the pollutant with the different clay-based materials reach values between 34% and 48% with a low consumption of H_2O_2 (20%–26%) with the natural, calcined and acid activated clays after 24 h of reaction and same operational conditions. The TOC abatement reached similar conversions (23%–29% after 24 h of reaction time) to those found for H_2O_2. It is also possible to observe that the mineralization reached with the clay-based materials is higher when compared to the non-catalytic run (9% after 24 h of reaction). Slight differences of the catalytic activity in the CWPO of PCM can be found for the non-pillared clays. The lowest catalytic activity was found for the calcined sample (KO-C), which can be explained by the lowest BET surface area of this material and the highest pH_{PZC} value [44]. The higher conversions of PCM, H_2O_2 and TOC obtained with the KO-A sample were ascribed to the highest values of acidity and basicity, as well as to its textural properties when compared to the other samples. Regarding the TOC removal results given in Figure 4c, it is possible to observe in the right axis that the oxidized intermediates produced may be refractory, since the TOC derived from PCM (TOC contribution determined as the subtraction of the theoretical TOC contribution of PCM from the measured TOC), increased continuously during the time evaluated in the experiments. Catalytic performance should be improved in order to obtain the complete removal of the model pollutant and, in addition, a catalyst able to remove some of the oxidized intermediates products. This endeavor was the main target that justified the preparation of the pillared catalysts.

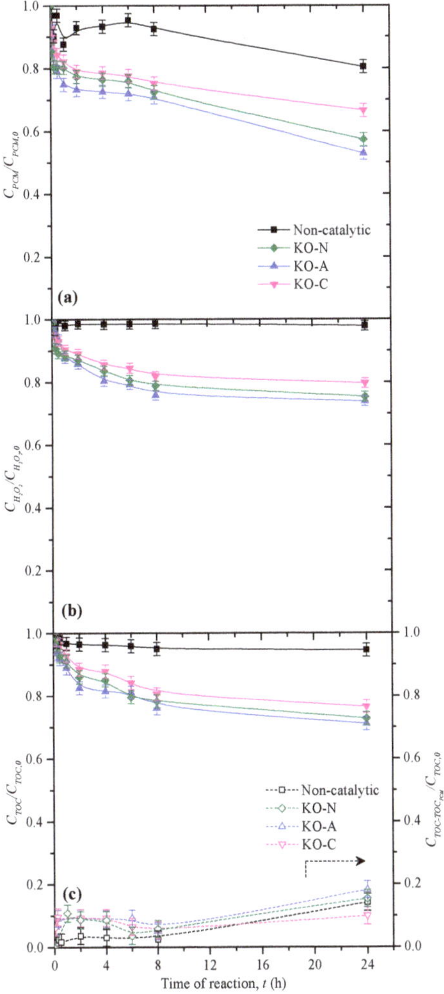

Figure 4. Profiles of the relative concentrations of (**a**) paracetamol (PCM), (**b**) H_2O_2 and (**c**) TOC during the catalytic wet peroxide oxidation (CWPO) of PCM with the non-pillared clays prepared from the natural clay extracted from the Kokshetau deposit. Operating conditions: C_{PCM} = 100 mg L^{-1}, $C_{H_2O_2}$ = 472 mg L^{-1}, C_{cat} = 2.5 g L^{-1}, initial pH = 3.5 and T = 80 °C.

The results of the CWPO of PCM obtained with the pillared clays prepared from the four different deposits considered in this work are represented in Figure 5. As observed, all pillared materials show higher catalytic activity when compared to the non-pillared Kokshetau samples previously shown. In fact, a significant difference can be observed between the pillared sample (KO-PILC) and the non-pillared samples, since the KO-PILC leads to a complete removal of PCM after 6 h. In addition, a mineralization of 79% is achieved with KO-PILC, whereas TOC abatement reached only 29% with the non-pillared samples. The highest catalytic activity of the KO-PILC in the CWPO of PCM when compared to the other Kokshetau samples can be explained only as a consequence of the presence of cobalt and iron, since the acidic and textural properties are similar to the non-pillared samples, evidencing that those metals are working as the main active phases in the catalyst. As can also be observed, all pillared materials present higher conversions of PCM, H_2O_2 and TOC (more than 78%, 71% and 28% after 8 h of reaction time, respectively) when compared to the non-pillared samples (34%, 20% and 23% at the same operating conditions, respectively). Two pillared samples (KO-PILC and KA-PILC) allow to remove completely the PCM after 6 h of reaction. Among them, KO-PILC show the highest catalytic activity in the CWPO of PCM, leading to the highest removal of TOC (68% after 8 h of reaction time). This can be ascribed to the highest BET surface area and to the highest amount of acidic and basic functionalities of the material, as well as to the highest content of iron (9%) of the natural clays, as observed in a previous work [21].

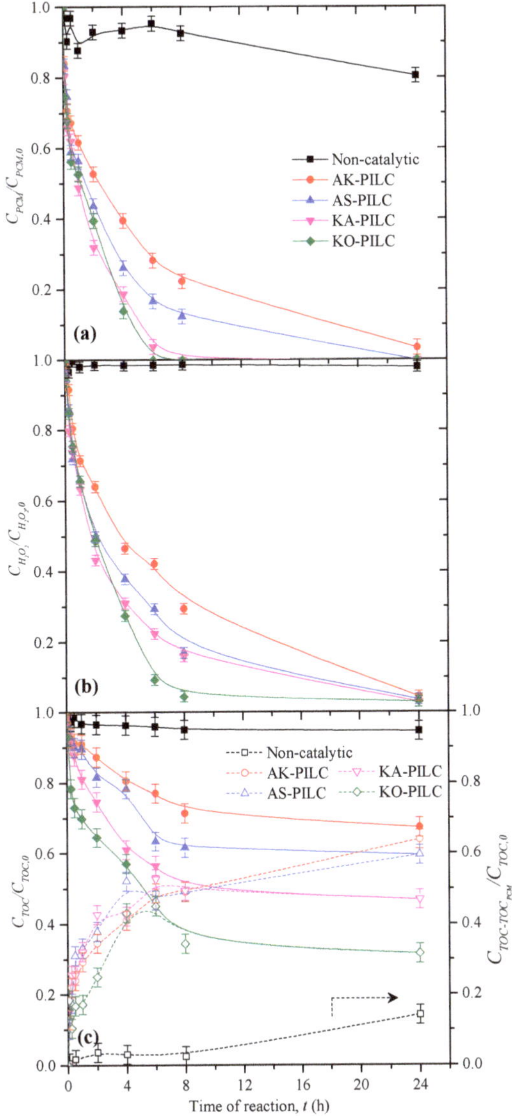

Figure 5. Profile of relative concentrations of (**a**) PCM, (**b**) H_2O_2 and (**c**) TOC during the CWPO of PCM with pillared clays. Operating conditions: C_{PCM} = 100 mg L^{-1}, $C_{H_2O_2}$ = 472 mg L^{-1}, C_{cat} = 2.5 g L^{-1}, initial pH = 3.5 and T = 80 °C.

As can be seen, the TOC profile free of the TOC contribution of paracetamol (expressed as $C_{TOC\text{-}TOC_{PCM}}$ and represented as open symbols in Figure 5c) reaches a maximum value close to 5 h, evidencing that some of the oxidized intermediates are further oxidized during the CWPO of PCM. After 6 h, when PCM disappeared, the profile is aligned with the measured TOC values and a partially decrease of the oxidized intermediates is then observed. In this work, the analysis of the different aliquots taken from the reaction media were done by HPLC, as described in the methodology section. However, calibrated compounds (*p*-nitrophenol, *p*-nitrocatechol, hydroquinone, *p*-benzoquinone, resorcinol, pyrocatechol, phenol and *trans*, *trans*-muconic acid) were not identified during the CWPO experiments, but a decrease of the pH during the experiments of CWPO was always observed, reaching values of 3.49, 3.35, 3.21, 3.18, 3.07, 3.00, 2.85 and 2.68 after 24 h in the non-catalytic run with the KO-C, KO-N, KO-A, AK-P, AS-P, KA-P and KO-P, respectively. This evidence supports the formation of carboxylic acid groups as oxidized intermediate compounds during the CWPO of PCM experiments. The material with highest performance (KO-PILC) lead to the minimal pH (2.68) observed after 24 h, meaning that a high extent of oxidation was reached.

To the best of our knowledge, there is no other work that uses modified clays in the CWPO of PCM, but it is possible to find studies regarding the degradation of PCM by Fenton-like process [20,45,46]. Velichkova et al. [20] obtained the complete removal of PCM after 4 h using a nanostructured maghemite powder catalyst under the following operating conditions: C_{PCM} = 100 mg L^{-1}, $C_{H_2O_2}$ = 28 mmol L^{-1}, C_{cat} = 6 g L^{-1}, initial pH = 2.6 and T = 60 °C. Similar results have been achieved in our work, since a complete removal of PCM is achieved using the lowest quantities of catalyst and hydrogen peroxide, despite a slightly higher temperature. Similar values of mineralization were also found in this work. Alalm et al. [45] achieved the complete removal of PCM after 1 h by photo-Fenton ($FeSO_4{\cdot}7H_2O$) at the following operating conditions: C_{PCM} = 100 mg L^{-1}, $C_{H_2O_2}$ = 1500 mg L^{-1}, C_{cat} = 0.5 g L^{-1} and initial pH = 3.0. In this case, PCM was removed quickly than in our study, but using an additional support of energy (the UV light) and adding a quantity of H_2O_2 considerably higher than the stoichiometric amount used in our work. In addition, the use of a homogeneous catalyst requires an additional process to recover the catalyst. In this previous study, the TOC or identified oxidized intermediates was not followed. Trovó et al. [46] also studied the photo-Fenton ($FeSO_4{\cdot}7H_2O$) of PCM (C_{PCM} = 50 mg L^{-1}, $C_{H_2O_2}$ = 120 mg L^{-1}, C_{cat} = 0.05 mM and initial pH = 2.5). Under those conditions, PCM was completely removed after 2 h. However, an additional source of energy and a homogeneous catalyst was used. A mineralization of 79% was achieved after 5 h of reaction, but the oxidized intermediates were not studied.

The adsorption and the homogeneous catalytic contribution in the removal of PCM was addressed by pure adsorption runs on the clays and by the determination of the leaching of iron from the materials during the CWPO experiments (Figure 6). PCM removal achieved in pure adsorption runs reached values from 4% to 18% after 24 h of contact time, values considerably lower when compared to the conversion of PCM obtained in the CWPO runs after 8 h of reaction, as also represented in Figure 6. This evidences that PCM is disappeared during CWPO runs because of the oxidation, instead of adsorption, since its contribution is poor in comparison with oxidation.

The concentration of iron determined after the experiments of CWPO was found to be lower than the limit concentration of 2 mg L^{-1} of iron in water courses, established by EU directives for treated water to be discharged into natural receiving water bodies. Curiously, the largest values of leached iron concentration were found with the natural and with the calcined materials (KO-N and KO-C, respectively). Iron coming from these materials is due to the iron content that is presented in the pristine material (9%) [21]. Among the non-pillared materials, the acid activated clay (KO-A) does not show leached iron precisely because of the treatment with the acid. Considering the values of the leaching of iron obtained with the non-pillared clays it is possible to consider that the homogeneous contributions are negligible. In fact, the maximum value of leached iron concentration among pillared clays was found with AK-PILC (0.097 mg L^{-1}) that showed the lowest catalytic activity in the CWPO of PCM. In addition, taking into account the theoretical maximum value of iron that is possible to observe in the media of reaction from the iron incorporated during the pillaring process (more than 800 mg L^{-1}), it is also possible to conclude that the materials show high stability (less than 0.015% of the iron present in the clays was leached).

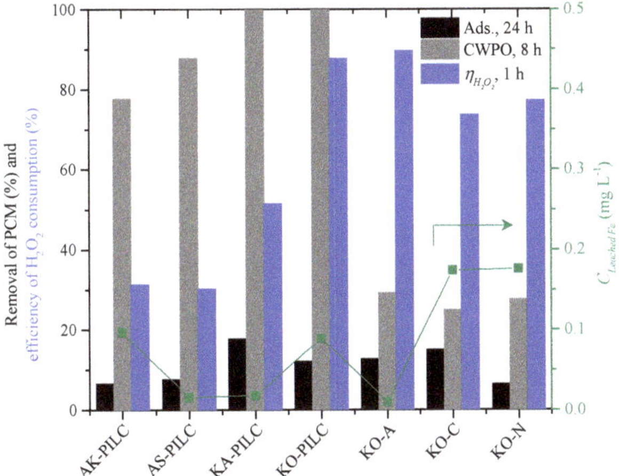

Figure 6. Pollutant removal by CWPO and by pure adsorption after 8 and 24 h, respectively; and H_2O_2 consumption efficiency and leached iron after 1 and 24 h of time of reaction, respectively in CWPO runs. Operating conditions: C_{PCM} = 100 mg L^{-1}, $C_{H_2O_2}$ = 472 mg L^{-1}, C_{cat} = 2.5 g L^{-1}, initial pH = 3.5 and T = 80 °C.

3. Materials and Methods

3.1. Reactants and Materials

Natural clays were supplied from four different deposits of Kazakhstan, viz. Akzhar, Asa, Karatau and Kokshetau (AK-N, AS-N, KA-N and KO-N samples, respectively). Acetic acid glacial (99.8%) and sodium acetate (98%) were obtained from Fischer Chemical and used in the buffer solution prepared for washing the clays. Iron (III) chloride hexahydrate (99%) and cobalt (II) chloride hexahydrate (99%) supplied from Aldrich and Fischer Chemical respectively, were used to prepare the pillaring solution. 4-acetamidophenol (paracetamol, 98%), obtained from Alfa Aesar, was used as model pollutant. Hydrogen peroxide (30% *w*/*v*), supplied by Fisher Chemical, was used as an oxidant. For analytical techniques, titanium (IV) oxysulfate (99.99%), sulfuric acid (98%), anhydrous sodium sulphite (98%), ortho-phosphoric acid (85%), acetonitrile (99.95%) and iron (II) chloride tetrahydrate (99%), obtained from Aldrich, Labkem, Panreac, Riedel-de Haen, VWR and Sigma-Aldrich respectively, were used. Ultrapure water was used in the preparation of solutions.

3.2. Preparation of Clay-Based Materials

Pillared clays were prepared using a pillaring solution that was prepared by dropwise addition of 0.5 M NaOH solution, at room temperature, to an aqueous solution of 0.5 M $FeCl_3$ and 0.25 M $CoCl_2$ to obtain a final solution with a molar ratio OH/(Fe + Co) = 2:1. Then, the pillaring solution was added into a 2 wt % natural clay suspension (natural clays were washed previously with a sodium acetate buffer solution). The resultant suspension was stirred at room temperature during 3 h, aged during 72 h and then the clay being recovered by filtration. The material was washed with water until the rinsing waters reach the natural pH, dried overnight at 60 °C and calcined at 600 °C during 5 h, resulting in AK-PILC, AS-PILC, KA-PILC, KO-PILC materials, obtained respectively from AK-N, AS-N, KA-N and KO-N samples. Additionally, one sample calcined at same conditions (600 °C for 5 h) was prepared from the KO-N in order to compare the effect of the calcination without the pillarization process, resulting in the sample KO-C. An acid activated clay was also prepared by immersing 3 g of KO-N in 150 mL of 4 M H_2SO_4 at 80 °C for 3 h to compare the effectiveness between pillared and acid-activated clays, resulting in the sample KO-A.

3.3. Techniques of Characterization

The textural properties of the materials were determined from N_2 adsorption–desorption isotherms at 77 K, obtained in a Quantachrome instrument NOVA TOUCH LX4 [47]. The specific surface area (S_{BET}) was calculated using the BET method using the Quantachrome TouchWin™ software 1.21. The total pore volume (V_{Total}) was considered at $p/p^0 = 0.98$. Fourier Transform Infrared (FT-IR) spectroscopy was performed with a Perkin Elmer FT-IR spectrophotometer UATR Two with a resolution of 1 cm^{-1} and scan range of 4000 to 450 cm^{-1} using the sample in powder solid state without further preparation. X-ray diffraction (XRD) analysis was performed in a PANalyticalX'Pert PRO equipped with a X'Celerator detector and secondary monochromator (Cu Kα λ = 0.154 nm; data recorded at a 0.017 degree step size). The crystallographic phases present were identified using a DEMO version of the HighScore software and the Crystallography Open Database.

The pH of the point of zero charge (pH_{PZC}) was determined by pH drift tests, as described elsewhere [44]. Briefly, five NaCl (0.01 M) solutions were prepared as electrolyte with varying initial pH (in the range 2–10, using HCl and NaOH 0.1 M solutions). Samples of 0.05 g of pillared clays were contacted with 20 mL of each NaCl solution. The equilibrium pH of each suspension was measured after 48 h under stirring (320 rpm) at room temperature. The pH_{PZC} value was determined by intercepting the curve 'final pH vs initial pH' with the straight line 'final pH = initial pH'.

The concentrations of acidic and basic sites of the clays were determined following the methodology described in a previous work for carbon-based materials [44]. Briefly, the concentration of acidic sites was determined by adding 0.2 g of each clay to 25 mL of a 0.02 mol L^{-1} NaOH solution. The resulting suspensions were left under stirring for 48 h at room temperature. After filtration, to remove the solid material, the unreacted OH^- was titrated with a 0.02 mol L^{-1} HCl solution. The initial concentration of acidic functionalities was then calculated by the difference between the amount of NaOH initially present in the suspension and the amount of NaOH determined by titration and dividing this value by the mass of material. The concentration of basic sites was determined in a similar way, this time by adding the carbon sample to a 0.02 mol L^{-1} HCl solution and titration with a 0.02 mol L^{-1} NaOH solution. Phenolphthalein was used as indicator in both titrations.

3.4. Oxidation Runs

Batch oxidation runs were carried out in a 250 mL well stirred round flask reactor equipped with a condenser and a temperature measurement thermocouple. An initial concentration of PCM of 100 mg L^{-1} was considered to model wastewaters containing pharmaceutical compounds. The reactor was loaded with 100 mL of the PCM aqueous solution and heated by immersion in an oil bath at controlled temperature. Upon stabilization at the desired temperature (80 °C), the solution pH was adjusted to a previously chosen value by means of H_2SO_4 solutions, and the experiments were allowed to proceed freely (not buffered). Then, the adequate quantity of 30% *w/v* H_2O_2 solution was added in order to use the stoichiometric dosage of H_2O_2 needed for PCM mineralization. Finally, the selected amount of catalyst was loaded (2.5 g L^{-1}), being that moment considered as the initial reaction time, $t_0 = 0$ h. All runs were conducted during 24 h. Pure adsorption runs were performed at the same operating conditions in the absence of H_2O_2, in order to compare with the pollutant removal obtained by CWPO experiments. Additionally, a blank experiment, in the absence of catalyst, was also carried out to observe the non-catalytic contribution to the drug degradation.

3.5. Analytical Techniques

Small aliquots were periodically withdrawn from the reactor, in order to be analysed by HPLC, TOC analysis and UV-Vis spectrophotometry, adapting methodologies described elsewhere [21,48]. PCM and its expected oxidized intermediate products (*p*-nitrophenol, *p*-nitrocatechol, hydroquinone, *p*-benzoquinone, resorcinol, pyrocatechol, phenol and *trans*, *trans*-muconic acid) were followed by using a Jasco HPLC system at a wavelength of 277 nm (UV-2075 Plus detector). For this purpose, a Kromasil

100-5-C18 column and 0.65 mL min^{-1} (PU-2089 Plus) of an A:B (10:90) mixture of acetonitrile (A) and sulfuric acid (pH = 3) aqueous solution (B) were used. The concentration of H_2O_2 was determined by adding the aliquot into a 5 mL volumetric flask containing 1 mL of H_2SO_4 solution (0.5 mol L^{-1}) and 0.1 mL of $TiOSO_4$. The resulting mixture was diluted with distilled water and further analyzed at 405 nm using a T70 spectrometer of PG Instruments Ltd. (Lutterworth, United Kingdom). Total organic carbon (TOC) was determined using a TOC-L CSN analyser of Shimadzu (Kyoto, Japan).

Leached iron was determined only for the last sample, withdrawn from the reaction media, by atomic absorption spectroscopy (Varian SpectrAA 220).

4. Conclusions

The preparation of clay-based materials by acid activation and by pillarization leads to an increase of the acidity character and of the specific surface area with respect to the corresponding natural clays. Despite this, the textural properties and the acidic functionalities were not found to affect significantly the catalytic activity of the clay materials in the CWPO of PCM, in opposition to the presence of Fe and Co in the prepared pillared clays. Whereas the maximum conversion of PCM achieved with non-pillared clays was 35% after 24 h, a complete removal of PCM was achieved with the Fe/Co-pillared clays under following conditions: C_{PCM} = 100 mg L^{-1}, $C_{H_2O_2}$ = 472 mg L^{-1}, C_{cat} = 2.5 g L^{-1}, initial pH = 3.5 and T = 80 °C. The acid activated clays presented a higher specific surface area and similar acidity character when compared to the pillared clays, but less catalytic activity in the CWPO process, supporting the conclusion that Fe and Co are responsible for the high activity shown by the pillared clays. At the tested conditions, the iron anchored in the clays as pillars was found to be stable since non-significant leaching of the iron was observed.

Author Contributions: Conceptualization: A.S.S., J.L.D.d.T. and H.T.G.; investigation: A.S.S. and M.S.K.; writing—original draft preparation: A.S.S. and J.L.D.d.T.; writing—review: H.T.G.; supervision: J.L.D.d.T. and H.T.G.; resources: H.T.G., J.L.D.d.T., M.S.K. and B.K.M.

Funding: This work is a result of the Project "AIProcMat@N2020—Advanced Industrial Processes and Materials for a Sustainable Northern Region of Portugal 2020", with the reference NORTE-01-0145-FEDER-000006, supported by Norte Portugal Regional Operational Programme (NORTE 2020), under the Portugal 2020 Partnership Agreement, through the European Regional Development Fund (ERDF); the Associate Laboratory LSRE-LCM—UID/EQU/50020/2019—funded by national funds through FCT/MCTES (PIDDAC); and CIMO (UID/AGR/00690/2019) through FEDER under Program PT2020.

Conflicts of Interest: The authors declare no conflict of interest.

References

1. Luo, Y.; Guo, W.; Ngo, H.H.; Nghiem, L.D.; Hai, F.I.; Zhang, J.; Liang, S.; Wang, X.C. A review on the occurrence of micropollutants in the aquatic environment and their fate and removal during wastewater treatment. *Sci. Total Environ.* **2014**, *473*, 619–641. [CrossRef]
2. Verlicchi, P.; Al Aukidy, M.; Zambello, E. Occurrence of pharmaceutical compounds in urban wastewater: Removal, mass load and environmental risk after a secondary treatment—A review. *Sci. Total. Environ.* **2012**, *429*, 123–155. [CrossRef] [PubMed]
3. Petrie, B.; Barden, R.; Kasprzyk-Hordern, B. A review on emerging contaminants in wastewaters and the environment: Current knowledge, understudied areas and recommendations for future monitoring. *Water Res.* **2015**, *72*, 3–27. [CrossRef] [PubMed]
4. Ratola, N.; Cincinelli, A.; Alves, A.; Katsoyiannis, A. Occurrence of organic microcontaminants in the wastewater treatment process. A mini review. *J. Hazard. Mater.* **2012**, *239*, 1–18. [CrossRef]
5. Kidd, K.A.; Blanchfield, P.J.; Mills, K.H.; Palace, V.P.; Evans, R.E.; Lazorchak, J.M.; Flick, R.W. Collapse of a fish population after exposure to a synthetic estrogen. *Proc. Natl. Acad. Sci. USA* **2007**, *104*, 8897–8901. [CrossRef]
6. Escher, B.I.; Baumgartner, R.; Koller, M.; Treyer, K.; Lienert, J.; McArdell, C.S. Environmental toxicology and risk assessment of pharmaceuticals from hospital wastewater. *Water Res.* **2011**, *45*, 75–92. [CrossRef]

7. Gadipelly, C.; Pérez-González, A.; Yadav, G.D.; Ortiz, I.; Ibáñez, R.; Rathod, V.K.; Marathe, K.V. Pharmaceutical industry wastewater: Review of the technologies for water treatment and reuse. *Ind. Eng. Chem. Res.* **2014**, *53*, 11571–11592. [CrossRef]
8. Rodea-Palomares, I.; Gonzalez-Pleiter, M.; Gonzalo, S.; Rosal, R.; Leganes, F.; Sabater, S.; Casellas, M.; Muñoz-Carpena, R.; Fernández-Piñas, F. Hidden drivers of low-dose pharmaceutical pollutant mixtures revealed by the novel GSA-QHTS screening method. *Sci. Adv.* **2016**, *2*, e1601272. [CrossRef] [PubMed]
9. Christophoridis, C.; Nika, M.-C.; Aalizadeh, R.; Thomaidis, N.S. Ozonation of ranitidine: Effect of experimental parameters and identification of transformation products. *Sci. Total Environ.* **2016**, *557*, 170–182. [CrossRef] [PubMed]
10. European Commission. Directive 2013/39/EU of the European Parliament and of the Council of 12 August 2013 Amending Directives 2000/60/EC and 2008/105/EC as Regards Priority Substances in the Field of Water Policy. *Off. J. Eur. Union* **2013**, *OJL 226*, 1–17.
11. European Commission. Decision (EU) 2015-495 of 20 March 2015 establishing a watch list of substances for Union-wide monitoring in the field of water policy pursuant to Directive 2008/105/EC of the European Parliament and of the Council. *Off. J. Eur. Union* **2015**, *OJL 78*, 40–42.
12. Villota, N.; Lomas, J.M.; Camarero, L.M. Kinetic modelling of water-color changes in a photo-Fenton system applied to oxidate paracetamol. *J. Photochem. Photobiol. A Chem.* **2018**, *356*, 573–579. [CrossRef]
13. Slamani, S.; Abdelmalek, F.; Ghezzar, M.R.; Addou, A. Initiation of Fenton process by plasma gliding arc discharge for the degradation of paracetamol in water. *J. Photochem. Photobiol. A Chem.* **2018**, *359*, 1–10. [CrossRef]
14. Akhi, Y.; Irani, M.; Olya, M.E. Simultaneous degradation of phenol and paracetamol using carbon/MWCNT/Fe_3O_4 composite nanofibers during photo-like-Fenton process. *J. Taiwan Inst. Chem. Eng.* **2016**, *63*, 327–335. [CrossRef]
15. Abdel-Wahab, A.-M.; Al-Shirbini, A.-S.; Mohamed, O.; Nasr, O. Photocatalytic degradation of paracetamol over magnetic flower-like TiO_2/Fe_2O_3 core-shell nanostructures. *J. Photochem. Photobiol. A Chem.* **2017**, *347*, 186–198. [CrossRef]
16. Augusto, T.D.M.; Chagas, P.; Sangiorge, D.L.; Leod, T.C.D.O.M.; Oliveira, L.C.; De Castro, C.S. Iron ore tailings as catalysts for oxidation of the drug paracetamol and dyes by heterogeneous Fenton. *J. Environ. Chem. Eng.* **2018**, *6*, 6545–6553. [CrossRef]
17. Silva, C.P.; Jaria, G.; Otero, M.; Esteves, V.I.; Calisto, V. Waste-based alternative adsorbents for the remediation of pharmaceutical contaminated waters: Has a step forward already been taken? *Bioresour. Technol.* **2018**, *250*, 888–901. [CrossRef]
18. Mirzaei, A.; Haghighat, F.; Chen, Z.; Yerushalmi, L. Removal of pharmaceuticals from water by homo/heterogonous Fenton-type processes—A review. *Chemosphere* **2017**, *174*, 665–688. [CrossRef]
19. Badawy, M.I.; Wahaab, R.A.; El-Kalliny, A. Fenton-biological treatment processes for the removal of some pharmaceuticals from industrial wastewater. *J. Hazard. Mater.* **2009**, *167*, 567–574. [CrossRef]
20. Velichkova, F.; Julcour-Lebigue, C.; Koumanova, B.; Delmas, H. Heterogeneous Fenton oxidation of paracetamol using iron oxide (nano)particles. *J. Environ. Chem. Eng.* **2013**, *1*, 1214–1222. [CrossRef]
21. Kalmakhanova, M.S.; Diaz de Tuesta, J.L.; Massalimova, B.K.; Gomes, H.T. Pillared clays from natural resources as catalysts for catalytic wet peroxide oxidation: Characterization and kinetic insights. *Environ. Eng. Res.* **2019**. [CrossRef]
22. Mnasri-Ghnimi, S.; Frini-Srasra, N. Catalytic wet peroxide oxidation of phenol over Ce-Zr-modified clays: Effect of the pillaring method. *Korean J. Chem. Eng.* **2015**, *32*, 68–73. [CrossRef]
23. Mnasri-Ghnimi, S.; Frini-Srasra, N. Effect of Al and Ce on Zr-pillared bentonite and their performance in catalytic oxidation of phenol. *Russ. J. Phys. Chem. A* **2016**, *90*, 1766–1773. [CrossRef]
24. Tomul, F.; Basoglu, F.T.; Canbay, H. Determination of adsorptive and catalytic properties of copper, silver and iron contain titanium-pillared bentonite for the removal bisphenol A from aqueous solution. *Appl. Surf. Sci.* **2016**, *360*, 579–593. [CrossRef]
25. Galeano, L.-A.; Vicente, M.A.; Gil, A. Catalytic degradation of organic pollutants in aqueous streams by mixed Al/M-pillared clays (M = Fe, Cu, Mn). *Catal. Rev.* **2014**, *56*, 239–287. [CrossRef]
26. Rhodes, C.N.; Franks, M.; Parkes, G.M.B.; Brown, D.R. The effect of acid treatment on the activity of clay supports for $ZnCl_2$ alkylation catalysts. *J. Chem. Soc. Chem. Commun.* **1991**, 804–807. [CrossRef]

27. Yu, W.; Wang, P.; Zhou, C.; Zhao, H.; Tong, D.; Zhang, H.; Yang, H.; Ji, S.; Wang, H. Acid-activated and WO -loaded montmorillonite catalysts and their catalytic behaviors in glycerol dehydration. *Chin. J. Catal.* **2017**, *38*, 1087–1100. [CrossRef]
28. Cool, P.; Vansant, E.F. Pillared Clays: Preparation, characterization and applications. In *Synthesis*; Springer: Berlin/Heidelberg, Germany, 2001; Volume 1, pp. 265–288.
29. Thommes, M.; Kaneko, K.; Neimark, A.V.; Olivier, J.P.; Rodríguez-Reinoso, F.; Rouquerol, J.; Sing, K.S. Physisorption of gases, with special reference to the evaluation of surface area and pore size distribution (IUPAC Technical Report). *Pure Appl. Chem.* **2015**, *87*, 1051–1069. [CrossRef]
30. Trigueiro, P.; Pereira, F.A.; Guillermin, D.; Rigaud, B.; Balme, S.; Janot, J.-M.; Dos Santos, I.M.; Fonseca, M.G.; Walter, P.; Jaber, M. When anthraquinone dyes meet pillared montmorillonite: Stability or fading upon exposure to light? *Dyes Pigments* **2018**, *159*, 384–394. [CrossRef]
31. Sprynskyy, M.; Sokol, H.; Rafińska, K.; Brzozowska, W.; Railean-Plugaru, V.; Pomastowski, P.; Buszewski, B. Preparation of AgNPs/saponite nanocomposites without reduction agents and study of its antibacterial activity. *Colloids Surf. B Biointerfaces* **2019**, *180*, 457–465. [CrossRef]
32. Zhu, B.-L.; Qi, C.-L.; Zhang, Y.-H.; Bisson, T.; Xu, Z.; Fan, Y.-J.; Sun, Z.-X. Synthesis, characterization and acid-base properties of kaolinite and metal (Fe, Mn, Co) doped kaolinite. *Appl. Clay Sci.* **2019**, *179*, 105138. [CrossRef]
33. Wu, C.; Wei, X.; Liu, P.; Tan, J.; Liao, C.; Wang, H.; Yin, L.; Zhou, W.; Cui, H.-J. Influence of structural Al species on Cd(II) capture by iron muscovite nanoparticles. *Chemosphere* **2019**, *226*, 907–914. [CrossRef]
34. Yuan, P.; Annabi-Bergaya, F.; Tao, Q.; Fan, M.; Liu, Z.; Zhu, J.; He, H.; Chen, T. A combined study by XRD, FTIR, TG and HRTEM on the structure of delaminated Fe-intercalated/pillared clay. *J. Colloid Interface Sci.* **2008**, *324*, 142–149. [CrossRef] [PubMed]
35. Liu, Y.; Dong, C.; Wei, H.; Yuan, W.; Li, K. Adsorption of levofloxacin onto an iron-pillared montmorillonite (clay mineral): Kinetics, equilibrium and mechanism. *Appl. Clay Sci.* **2015**, *118*, 301–307. [CrossRef]
36. Widjaya, R.R.; Juwono, A.L.; Rinaldi, N. Bentonite modification with pillarization method using metal stannum. *AIP Conf. Proc.* **2017**, *1904*, 020010.
37. Bahranowski, K.; Włodarczyk, W.; Wisła-Walsh, E.; Gaweł, A.; Matusik, J.; Klimek, A.; Gil, B.; Michalik-Zym, A.; Dula, R.; Socha, R.; et al. [Ti,Zr]-pillared montmorillonite—A new quality with respect to Ti- and Zr-pillared clays. *Microporous Mesoporous Mater.* **2015**, *202*, 155–164. [CrossRef]
38. Bruckman, V.J.; Wriessnig, K. Improved soil carbonate determination by FT-IR and X-ray analysis. *Environ. Chem. Lett.* **2013**, *11*, 65–70. [CrossRef]
39. Li, T.; Zhao, L.; Zheng, Z.; Zhang, M.; Sun, Y.; Tian, Q.; Zhang, S. Design and preparation acid-activated montmorillonite sustained-release drug delivery system for dexibuprofen in vitro and in vivo evaluations. *Appl. Clay Sci.* **2018**, *163*, 178–185. [CrossRef]
40. Jain, S.; Datta, M. Montmorillonite-alginate microspheres as a delivery vehicle for oral extended release of Venlafaxine hydrochloride. *J. Drug Deliv. Sci. Technol.* **2016**, *33*, 149–156. [CrossRef]
41. Komadel, P. Acid activated clays: Materials in continuous demand. *Appl. Clay Sci.* **2016**, *131*, 84–99. [CrossRef]
42. Eren, E.; Afsin, B. An investigation of Cu(II) adsorption by raw and acid-activated bentonite: A combined potentiometric, thermodynamic, XRD, IR, DTA study. *J. Hazard. Mater.* **2008**, *151*, 682–691. [CrossRef] [PubMed]
43. Wang, S.; Dong, Y.; He, M.; Chen, L.; Yu, X. Characterization of GMZ bentonite and its application in the adsorption of Pb(II) from aqueous solutions. *Appl. Clay Sci.* **2009**, *43*, 164–171. [CrossRef]
44. De Tuesta, J.D.; Quintanilla, A.; Casas, J.; Rodriguez, J. P-, B- and N-doped carbon black for the catalytic wet peroxide oxidation of phenol: Activity, stability and kinetic studies. *Catal. Commun.* **2017**, *102*, 131–135. [CrossRef]
45. Alalm, M.G.; Tawfik, A.; Ookawara, S. Degradation of four pharmaceuticals by solar photo-Fenton process: Kinetics and costs estimation. *J. Environ. Chem. Eng.* **2015**, *3*, 46–51. [CrossRef]
46. Trovó, A.G.; Nogueira, R.F.P.; Agüera, A.; Fernández-Alba, A.R.; Malato, S. Paracetamol degradation intermediates and toxicity during photo-Fenton treatment using different iron species. *Water Res.* **2012**, *46*, 5374–5380. [CrossRef] [PubMed]
47. De Tuesta, J.L.D.; Silva, A.M.; Faria, J.L.; Gomes, H.T. Removal of Sudan IV from a simulated biphasic oily wastewater by using lipophilic carbon adsorbents. *Chem. Eng. J.* **2018**, *347*, 963–971. [CrossRef]

48. Diaz de Tuesta, J.L.; Machado, B.F.; Serp, P.; Silva, A.M.T.; Faria, J.L.; Gomes, H.T. Janus amphiphilic carbon nanotubes as Pickering interfacial catalysts for the treatment of oily wastewater by selective oxidation with hydrogen peroxide. *Catal. Today* **2019**. [CrossRef]

Article

Magnetic Nanoparticles for Photocatalytic Ozonation of Organic Pollutants

Carla A. Orge *, O. Salomé G. P. Soares, Patrícia S. F. Ramalho, M. Fernando R. Pereira and Joaquim L. Faria

Laboratory of Separation and Reaction Engineering—Laboratory of Catalysis and Materials, (LSRE-LCM), Faculdade de Engenharia da Universidade do Porto, Rua Dr. Roberto Frias, 4200-465 Porto, Portugal
* Correspondence: carlaorge@fe.up.pt

Received: 30 July 2019; Accepted: 19 August 2019; Published: 22 August 2019

Abstract: Magnetic nanoparticles (MNP) composed of iron oxide (or other metal–FeO cores) coated with carbon produced by chemical vapour decomposition (CVD) were used in the photocatalytic ozonation of oxamic acid (OMA) which we selected as a model pollutant. The incorporation of Ag and Cu on FeO enhanced the efficiency of the process. The carbon phase significantly increased the photocatalytic activity towards the conversion of OMA. As for the synthesis process, raising the temperature of CVD improved the performance of the produced photocatalysts. The obtained results suggested that the carbon phase is directly related to high catalytic activity. The most active photocatalyst (C@FeO_CVD850) was used in the removal of other compounds (dyes, industrial pollutants and herbicides) from water and high mineralization levels were attained. This material was also revealed to be stable during reutilisation.

Keywords: magnetic materials; chemical vapour deposition; photocatalytic ozonation; organic pollutants

---

1. Introduction

Photocatalytic ozonation (PCO) results from the combination of two different techniques, and consequently, the potential of pollutant abatement increases since the generation of hydroxyl radicals is promoted [1]. According to the literature, besides direct ozonation in the presence of suitable catalysts under illumination, O_3 can generate hydroxyl radicals ($HO^\bullet$) through the formation of ozonide radical ($O_3{}^{\bullet-}$) in the adsorption layer of the optical semiconductor (SC) where surface holes ($h^+_{[SC]}$) and electrons ($e^-_{[SC]}$) are generated, as described in Equations (1), (2) and (3). The generated $O_3{}^{\bullet-}$ species rapidly reacts with H^+ to produce the radical $HO_3{}^\bullet$, which evolves to O_2 and $HO^\bullet$. The generated $O_3{}^{\bullet-}$ species quickly reacts with H^+ in the solution to give the $HO_3{}^\bullet$ radical (Equation (4)), which results in O_2 and $HO^\bullet$ (Equation (5)).

$$[SC] + h\nu \rightarrow h^+_{[SC]} + e^-_{[SC]} \tag{1}$$

$$O_3 + e^-_{[SC]} \rightarrow O_3^{\bullet-} \tag{2}$$

$$h^+_{[SC]} + H_2O \rightarrow HO^\bullet + H^+ \tag{3}$$

$$O_3^{\bullet-} + H^+ \rightleftarrows HO_3^\bullet \tag{4}$$

$$HO_3^\bullet \rightarrow HO^\bullet + O_2 \tag{5}$$

The reaction between O_3 and electrons on the catalyst surface interferes with the recombination of electrons and positive holes. Consequently, a more significant number of radicals are produced, thereby accelerating the photocatalytic reaction. Several studies have confirmed the synergistic effects of PCO on the degradation and removal of different substances [1,2].

Examining the papers published in recent years, most of the studies reported the use of AEROXIDE© TiO_2 P25 (Evonik), the "gold standard" in photocatalysis [3], during photocatalytic ozonation. On the other hand, magnetic nanoparticles (MNP) gained increasing importance in different areas such as biomedicine/biotechnology [4–6], catalysis [7–9] and water treatment [10–16]. In spite of being widely used as catalyst supports, carbon materials as a catalyst are attracting a great deal of attention [17]. Coating MNP with a layer of different materials improves their stability and introduces new surface properties and functionalities. The use of MNP coated with carbon will enable catalyst recovery by magnetic separation and also take advantage of the catalytic properties of the carbon materials. Magnetic separation to remove the catalyst from liquid solutions eliminates the necessity of centrifugation, filtration or other impractical techniques for recycling the catalysts, avoiding their agglomeration and consequently increasing their durability [9].

Although the use of MNP has been extensively reported, the development of new catalysts with magnetic properties for application in a wide range of expertise fields is still a challenge. In the case of advanced oxidation processes (AOPs), the use of MNP has also been a subject of intense research, but few works reported the presence of MNP during PCO. Mahmoodi used $CuFe_2O_4$ MNP to study the degradation of a dye in aqueous solution, and his results showed that these nanoparticles are very efficient in the degradation of the selected pollutant [14]. Mahmoodi et al. also investigated PCO for the removal of dyes using nickel–zinc ferrite (NZFMN) nanoparticles and 100% effective removal was verified [15]. Magnetic $ZnFe_2O_4$–C_3N_4 hybrids prepared by a simple reflux treatment were applied in photo-Fenton discolouration of Orange II [18]. Yin et al. [16] used $BiFeO_3$ magnetic nanoparticles as a visible light photocatalyst in the coupling of PCO for the degradation of oxalic acid and norfloxacin. A notable improvement allied to excellent stability during reuse reactions was confirmed. Recently, Hussan et al. [19] verified a high catalytic activity, stability and reuse of $BiFeO_3$ magnetic nanoparticles during aniline degradation. $CoFe_2O_4$ particles with a predominantly mesoporous structure containing a large surface area showed a higher decolourization and mineralization of melanoidin (aminocarbonyl complex polymers containing a dark brown colour present in several distillery wastewaters) during catalytic ozonation [20]. The improvement when compared to O_3 alone was attributed to the generation of $HO^{\bullet}$ radicals in the reaction medium. Magnetic Fe_2O_3 supported on ordered mesopours silica material (SBA-15) was synthesized by a facile impregnation and calcination method and the photocatalytic activities were also evaluated on the degradation of rhodamine B [21]. The structure and amount of catalyst and the concentration of H_2O_2 and the pollutant were the main influencing factors on the degradation of rhodamine B. The prepared magnetic Fe_2O_3/SBA-15 had excellent reusability and stability, showing only a slightly decrease after five consecutive runs. Catalytic ozonation of di-n-butyl phthalate (DBP) was carried out in the presence of Ag-doped $MnFe_2O_4$ catalysts [22]. The Ag-doped $MnFe_2O_4$ enhanced the apparent rate constants compared to O_3 alone and undoped $MnFe_2O_4$ systems, and the catalyst surface hydroxyl groups were identified as a critical factor. The Bi_2WO_6–FeOx photocatalysts presented with stable photocatalytic activity during repetitive norfloxacin degradation experiments [22]. Ortiz-Quiñonez et al. showed that MNP containing Ni prepared by a low-temperature solution combustion method were very active catalysts for the degradation of 4-nitrophenol in aqueous media [23]. Magnetic Fe_3O_4/SnO_2 nanocomposites with different molar ratios were tested in the photodegradation of crystal violet [24]. The photoconductivity studies revealed the ohmic nature of the samples was responsible for the good photocatalytic activity due to the extended photoresponsive range and increase in charge separation rate. A novel rattle-type magnetic Fe_3O_4@Ag@H–BiOCl nanocomposite was successfully prepared by a solvothermal method and evaluated by the photocatalytic degradation of rhodamine B and the antibiotic ciprofloxacin [25]. The characterization results showed that the composites exhibit an obvious cavity that promotes excellent adsorption, showing better performance than pure BiOCl.

The aim of this work is to evaluate the PCO performance of MNP composed of iron oxide and coated with carbon by chemical vapour decomposition (CVD). Oxamic acid (OMA) was selected as a model compound since it is a simple molecule and one of the common final products that result from the degradation of a wide range of pollutants with high refractory character to oxidation [26]. Non-catalytic combined treatment (photo-ozonation), non-catalytic individual processes (single ozonation and photolysis), and the corresponding catalytic individual processes (catalytic ozonation and photocatalysis) were also carried out to understand the reaction mechanisms and evaluate the presence of synergetic effects during OMA removal by PCO. The catalytic activity of the prepared materials was confirmed in the degradation of other pollutants (Colour Index Reactive Blue Dye 5—DYE; aniline—ANL; metolachlor—MTLC) using the most active photocatalysts.

2. Results

2.1. Catalyst Characterisation

Table 1 displays the Brunauer–Emmet–Teller (BET) surface area and the amount of carbon (C) of the prepared samples. The textural characterisation of MNP revealed that the incorporation of Cu or Co in Fe_3O_4 (sample FeO) during the synthesis of these samples increased the BET surface area in contrast with the introduction of Ag that slightly decreased this value. This decrease may rely on the use of the incipient wetness impregnation method to deposit Ag on the FeO surface. In the case of MNP coated with carbon by CVD, we calculated the surface area per gram of carbon previously determined by thermogravimetric (TG) analysis. A drastic decrease to values lower than 10 $m^2\ g^{-1}$ was observed in the surface area of MNP coated by CVD. This decrease is due to the heat treatments used during synthesis, namely the heating up to 400 °C to carry out the reduction of the metal under hydrogen and, after that, a temperature increase up to 750 °C or 850 °C to coat the MNP with carbon [13].

Table 1. Brunauer–Emmet–Teller (BET) surface area (S_{BET}) and amount of carbon. CVD = chemical vapour decomposition.

Sample	S_{BET} ($m^2\ g^{-1}$) (±10 $m^2\ g^{-1}$)	% C †
FeO	154	0
CoFeO	184	0
Cu FeO	235	0
AgFeO	134	0
C@FeO_CVD750	63 *	16
C@FeO_CVD850	29 *	35
C@CoFeO_CVD750	29 *	34
C@CoFeO_CVD850	33 *	43
C@CuFeO_CVD750	27 *	67
C@CuFeO_CVD850	16 *	63
C@TiFeO_met1_CVD850	25 *	20
C@TiFeO_met2_CVD850	26 *	33

* per gram of carbon. † determined by thermogravimetric (TG) analysis.

In general, materials prepared by CVD at 850 °C had a higher amount of carbon than those at 750 °C.

In the transmission electron microscopy (TEM) of selected samples (Figure 1), it is possible to see that MNP are composed of small spheres that are not well defined and have no clear distinction of the different metals, which is indicative of a good dispersion of both metals. TEM images confirm the covering of the MNP with carbon during the CVD process. In the samples C@FeO_CVD750 and C@FeO_CVD850, we can confirm that iron oxide particles are encapsulated (coated with carbon nanotubes and nanofibers). The results suggest that the increase in temperature increased the formation of carbon, as confirmed by the TG analysis. In the case of C@CoFeO_CVD750, MNP are covered by few carbon layers and also by carbon nanotubes and nanofibers, while C@CoFeO_CVD850 appeared to be encapsulated by carbon nanotubes and nanofibers. Most of the MNP were encapsulated by carbon, but some metal oxide is exposed on the surface.

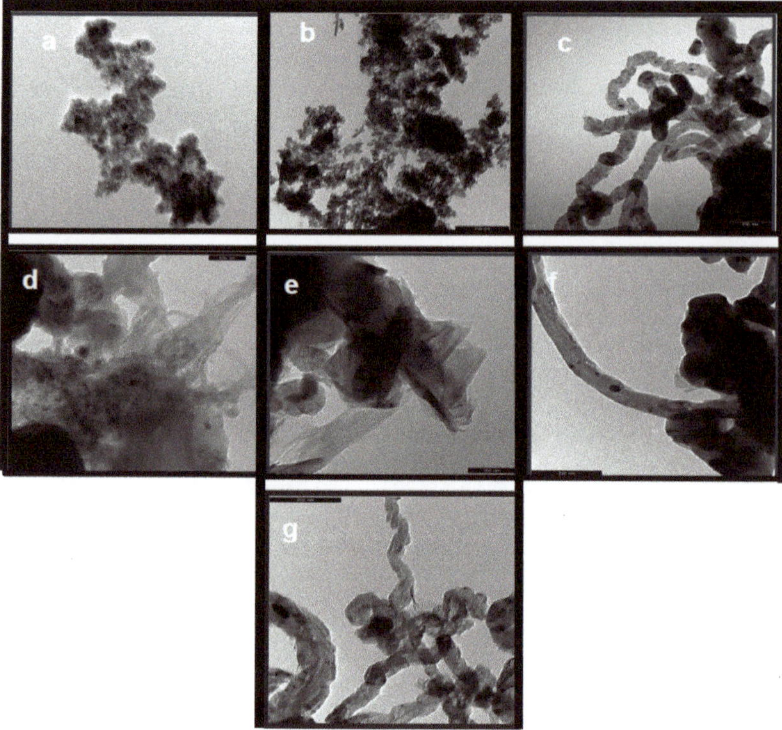

Figure 1. Transmission electron microscopy (TEM) micrographs of selected samples: (**a**) FeO; (**b**) AgFeO; (**c**) C@FeO_CVD750; (**d**) C@FeO_CVD850; (**e**) C@CoFeO_CVD750; (**f**) C@CoFeO_CVD850; (**g**) C@TiFeO_met2_CVD850.

The crystal structure of the prepared MNP was studied by X-ray diffraction (XRD) and the spectra of some of the tested catalysts are depicted in Figure 2. In Table 2, we show the identification of the phases identified in the MNP materials, along with the crystallite sizes. XRD analyses show that the only phase present in the FeO sample is magnetite. The main crystalline phase of AgFeO is also magnetite (92.6%). As expected, the coated MNP have a large amount of carbon in their composition, appearing as graphite with a percentage between 84% and 96%.

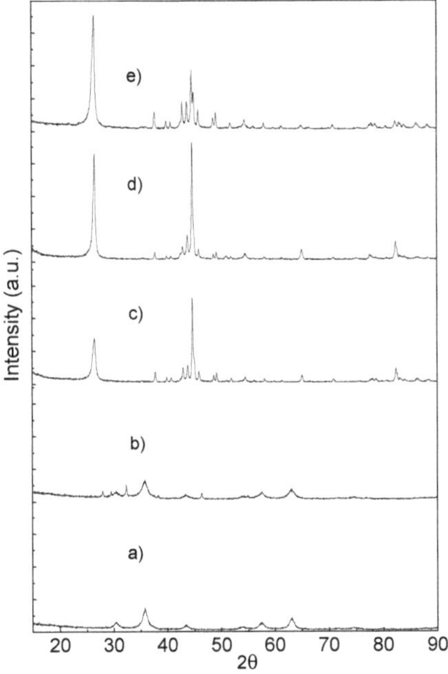

Figure 2. X-ray diffraction (XRD) spectra of selected samples: (**a**) FeO; (**b**) AgFeO; (**c**) C@FeO_CVD750; (**d**) C@CoFeO_CVD850; (**e**) C@TiFeO_met2_CVD850. a.u.: arbitrary units; 2θ: angle between transmitted beam and reflected beam.

Table 2. Properties of selected materials obtained by XRD analysis.

Sample	Phase (%vol/vol)	Crystallite Size (nm) ±6%
FeO	100% magnetite (Fe_3O_4)	20.5
$AgFe_3O_4$	7.3% chlorargyrite (AgCl) 92.6% magnetite (Fe_3O_4)	- 8.2
C@FeO_CVD750	3% cementite (Fe_3C) 1.6% Fe α 95.4% graphite	84 72 16
C@FeO_CVD850	3% cementite (Fe_3C) 1.5% Fe α 95.5% graphite	52 73 27
C@TiFeO_met2_CVD850	14.1% cementite (Fe_3C) 2.3% Fe α 83.6% graphite	72 10 -

2.2. OMA Removal by MNP

The conversion of OMA by PCO in the presence of different MNP is displayed in Figure 3 (ratio of pollutant concentration/initial pollutant concentration, C/C_0, in function of the time, t). The non-catalytic run was also performed, as well as the individual methods in the absence of catalyst, single ozonation and photolysis as control experiments.

All tested catalysts performed better than the non-catalytic run. The incorporation of Ag and Cu in FeO particles increased the catalytic activity, while the introduction of Co slightly decreased the performance. Although the introduction of Ag in FeO particles reduced the BET surface area, this sample had the highest OMA removal rate, suggesting that the addition of Ag leads to an increase in activity. Sampaio et al. [27] also observed increased efficiency in phenol degradation in the presence of ZnO loaded with Ag. The authors suggested that there is an optimal amount of Ag to generate an adequate quantity of holes and electrons, and an increase in the Ag load may lead to a surplus charge separation, which is not efficiently contributing to the reaction.

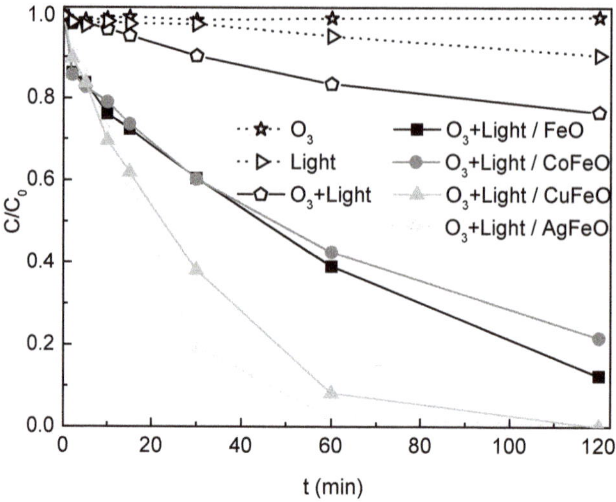

Figure 3. Conversion of oxamic acid (OMA) by photocatalytic ozonation (PCO) in the presence of magnetic nanoparticles (MNP).

2.3. OMA Removal by Carbon Coated MNP by Using 750 °C CVD

FeO, CuFeO and CoFeO particles were coated with carbon by CVD at 750 °C and tested in the removal of OMA by PCO. The conversions of OMA in the presence of several MNP observed up to 120 min of reaction are shown in Figure 4.

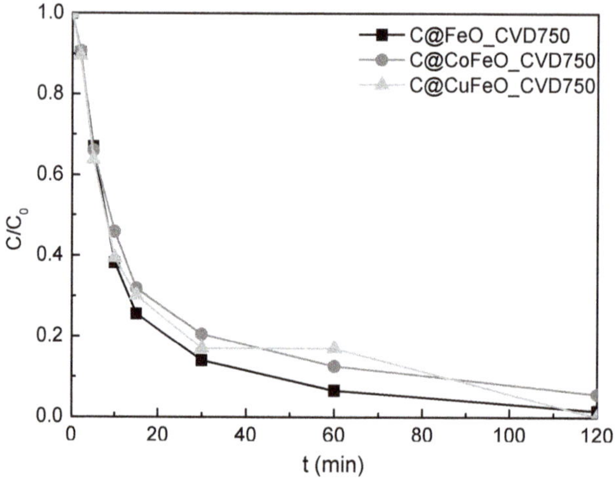

Figure 4. Conversion of OMA by PCO in the presence of MNP coated by CVD at 750 °C.

All MNP coated with carbon by CVD at 750 °C showed better performance than the corresponding uncoated MNP. This result suggests that the carbon phase plays a vital role during PCO, increasing the catalytic activity. However, there are no significant differences among the samples prepared by CVD at 750 °C (Figure 4). The introduction of Co or Cu in FeO particles followed by a CVD treatment at this temperature did not improve the OMA removal rate. The slight increase verified in the C/C_0 value at 60 min during the PCO with C@CuFeO_CVD750 is within the experimental error ($<\pm 2\%$). As observed in the TEM micrographs, MNP were encapsulated by carbon after the CVD treatment, suggesting that the carbon phase is mainly responsible for the improvement in activity. This result is in line with the fact that the presence of the carbon phase reduces the electron/hole recombination due to the formation of solid–solid interfaces that facilitate charge transfer and spatial separation, and improves interfacial defect sites that act as "hot spots" [28]. It is expected that the carbon phase, due to its electric conductivity, acts as photosensitizer to promote electric charge transfer, since it can enhance electron diffusion and reduce electron/hole recombination, and consequently the catalytic activity is improved.

2.4. OMA Removal by Carbon Coated MNP by Using 850 °C CVD

FeO, CoFeO and CuFeO coated with carbon by CVD at a higher temperature, 850 °C, were tested in the same degradation of OMA by PCO (Figure 5a). Because the catalysts are more active, the running time of the experiments was cut down to only 60 min, half of the time of the runs in the presence of the 750 °C CVD coated materials. The increase of the CVD temperature produced a positive effect on the catalytic performance during OMA removal. All MNP coated at 850 °C achieved complete OMA removal after 60 min of reaction. In general terms, the result correlates with the amount of deposited carbon to attain a specific coverage of the MNP. The limit value should lie around the 67% obtained for the Cu modified catalysts. Therefore, besides a thermal effect, the coating requires a certain amount of carbon to become more efficient. Temperatures above 850 °C were outside our experimental range, but there is no significant expected increase in carbon mass production above these temperatures, judging from experience in growing carbon nanotubes by CVD [29].

Taking advantage of the pronounced photoactivity of the Ti-modified materials, we used the TiFeO particles coated with carbon by CVD at 850 °C to test the efficiency of two different preparation methods of the bimetallic photocatalysts. In the first method, TiFeO particles were prepared by co-precipitation, as in the case of CuFeO and CoFeO, while in the second method, the particles were prepared by impregnation. The performances of these photocatalysts were compared to that of pure TiO_2 (synthesised by the sol-gel procedure according to [30]), and commercial TiO_2 (P25) (Figure 5b).

Independently of the preparation method, the prepared MNP containing Ti coated with carbon had a better efficiency than TiO_2 synthesised by the sol-gel procedure and more unsatisfactory performance than P25. The photocatalyst prepared by the second method revealed a slightly higher OMA removal rate than the one obtained by the first method. Again, the observation correlates with a higher amount of carbon.

It is worth noting that all the modified samples underperformed sample C@FeO_CVD850, suggesting that it is not worth introducing a new metal in FeO before coating by CVD. Similarly to samples treated by CVD at 750 °C, we encapsulated MNP with carbon nanotubes and nanofibers after the CVD process at 850 °C, suggesting that the carbon phase is mainly responsible for the efficiency rather than MNP composition.

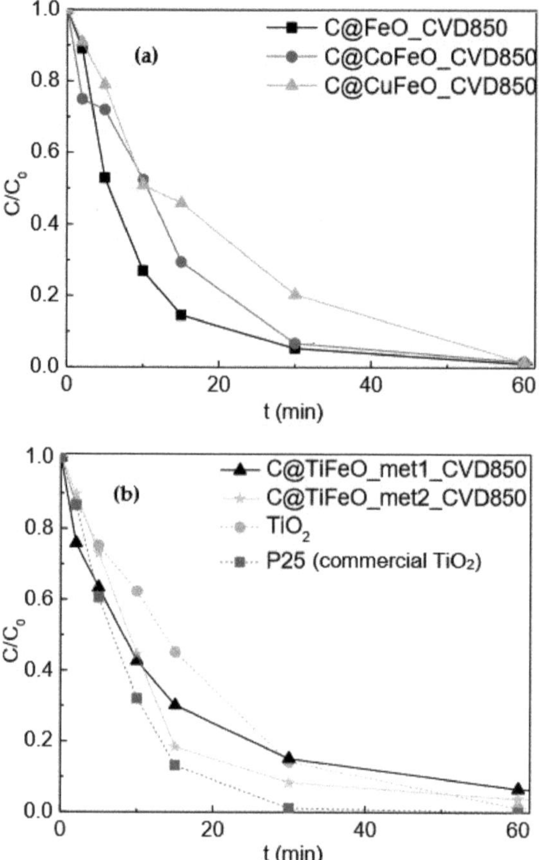

Figure 5. (**a**) OMA removal by PCO in the presence of MNP coated by CVD at 850 °C and (**b**) in the presence of Ti-based materials.

2.5. Testing the Removal Efficiency under Different Conditions

To better understand the processes evolving during OMA removal by PCO, additional experiments in the presence of the best catalyst, C@FeO_CVD850, were carried out. PCO resulting from the combination of two AOPs (ozonation and photocatalysis), potentiating the occurrence of synergetic effects. Thus, we compared the resulting levels of OMA removal in adsorption, catalytic ozonation and photocatalysis experiments performed under the same experimental conditions (Figure 6).

Adsorption on C@FeO_CVD850 was negligible. C@FeO_CVD850 was not a suitable catalyst for ozonation, removing only 10% of OMA after 60 min of reaction. The low catalytic activity verified during ozonation can be related to the small surface area of the C@FeO_CVD850 sample. On the other hand, the tested catalyst was photoactive, achieving a pollutant degradation of approximately 70% after 60 min of reaction. The verified catalytic activity probably resulted from the available electrons on the catalyst surface caused by the presence of metal oxide. However, the verified efficiency during PCO is higher than the sum of the performances obtained during photocatalysis and catalytic ozonation, suggesting the presence of a synergy when O_3, light and C@FeO_CVD850 were acting simultaneously. The high catalytic activity verified with the best sample resulted mainly due to the presence of the carbon phase. However, the presence of a metal oxide on the catalyst surface slightly contributed to the removal achieved by PCO. A more significant number of radicals is expected to generate when O_3, UV radiation and an appropriate catalyst are together, accounting for the high activity observed [30,31].

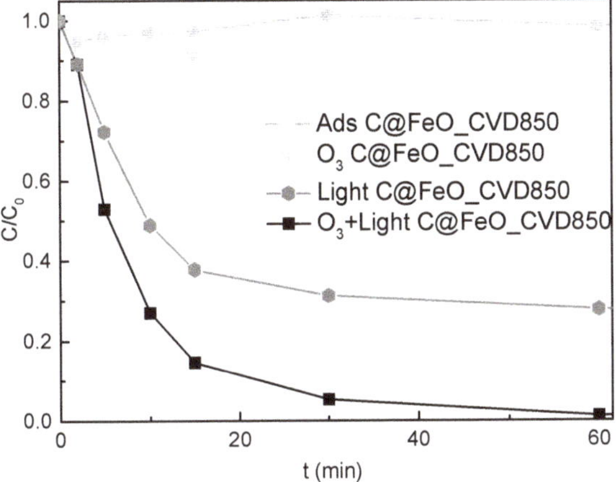

Figure 6. OMA removal by adsorption (Ads), catalytic ozonation, photocatalysis and PCO in the presence of C@FeO_CVD850.

The reutilisation of C@FeO_CVD850 was carried out to study the eventual deactivation during photocatalytic ozonation. For these experiments, a new solution was admitted to the reactor at the beginning of every reaction cycle (Figure 7).

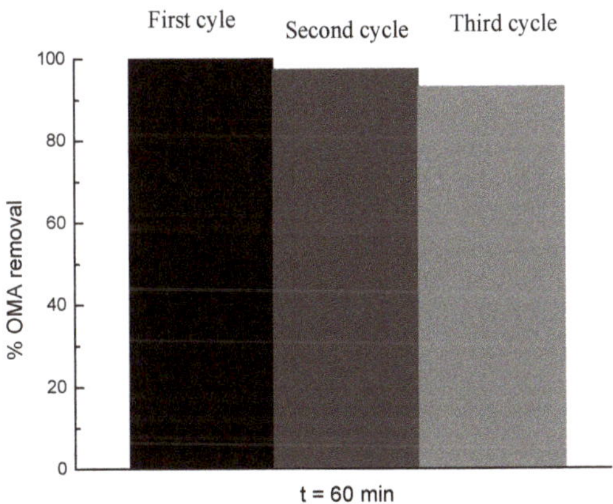

Figure 7. Runs of photocatalytic ozonation in the presence of C@FeO_CVD850.

Although there is a systematic decrease in the conversion of OMA after 60 min of reaction with C@FeO_CVD850, the photocatalyst remained active even after three cycles. A 7% loss in performance was observed from the first use to the third cycle of experiments (respectively 100%, 97% and 93% conversion measured in the consecutive runs). Since adsorption of OMA was demonstrated to be negligible, this is probably due to overoxidation of the surface by the accumulation of reactive oxygen species that will hinder the semiconductive optical properties of the material as well the active sites for ozone adsorption.

2.6. Removal of Other Pollutants by PCO

To assess the efficiency of PCO in the presence of C@FeO_CVD850 for the degradation of more complex compounds, we selected three representatives of different classes of pollutants namely: C.I. Reactive Blue 5 (DYE) as model of a textile effluent, aniline (ANL) as an example of an industrial pollutant and metolachlor (MTLC) as an emergent compound provided from the most widely used class of herbicides. Since direct ozonation is efficient to degrade compounds containing either an activated aromatic ring or double bonds, O_3 can remove these pollutants, but it presents limitations to achieve high mineralisation degrees [32,33]. The conversion of the selected contaminants over 30 min of PCO in the presence of C@FeO_CVD850 is represented in Figure 8a–c. As expected, the presence of O_3 was enough to remove the selected parent pollutants in a short reaction time. In the case of the DYE removal, the three processes had similar efficiencies. However, for ANL and MTLC, PCO was revealed to be more effective than single ozonation.

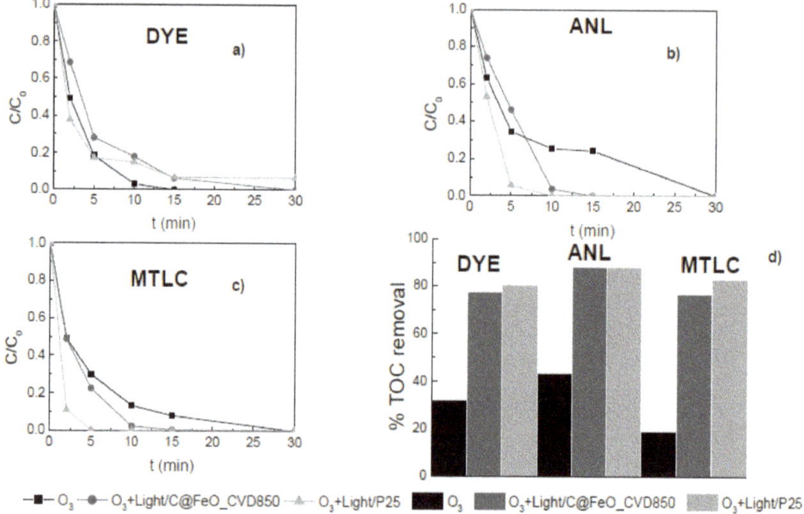

Figure 8. Evolution of dimensionless concentration of C.I. Reactive Blue 5 (DYE) (**a**), aniline (ANL) (**b**) and metolachlor (MTLC) (**c**) over 30 min, and the respective total organic carbon (TOC) removal after 180 min of reaction (**d**) by single ozonation and photocatalytic ozonation in the presence of C@FeO_CVD850 and P25.

We compared the degree of mineralisation measured in terms of total organic carbon (TOC) removal after 180 min for three standard processes: single ozonation, PCO in the presence of P25, and PCO in the presence of C@FeO_CVD850 (Figure 8d). In terms of mineralization level, the combination of O_3 with radiation and a catalyst was necessary to improve the organic matter removal. The prepared MNP coated with carbon by CVD at 850 °C showed a TOC removal between 78% and 88% after 180 min of reaction, suggesting a high rate of degradation and conversion into CO_2 and H_2O.

The photocatalyst C@FeO_CVD850 showed no lesser performance than the commercial benchmark P25 (TiO_2) in the PCO but had the additional advantage of possessing magnetic properties to facilitate separation from the aqueous solution.

3. Materials and Methods

3.1. Catalyst Preparation and Characterisation

The co-precipitation method was used to prepare FeO, CoFeO and CuFeO. Further details about FeO and CoFeO preparation can be found elsewhere [13]. For the synthesis of CuFeO, the method followed the CoFeO procedure, but replaced the cobalt precursor with $Cu(NO_3)_2{\cdot}3H_2O$.

AgFeO was prepared by incipient wetness by dropping $AgNO_3$ into the FeO support. The silver content was 5% wt. After that, the sample was dried at 100 °C for 24 h and then thermal treated under nitrogen flow at 200 °C for 2 h.

As Ti-based materials are highly active for PCO, two FeTiO samples were prepared. Sample FeTiO_m1 was obtained as CoFeO and CuFeO samples using $Ti[OCH(CH_3)_2]_4$ as a Ti precursor instead of Co and Cu precursors. Sample FeTiO_m2 was synthesised by dispersing FeO in alcohol for 1 h under ultrasonication [34], followed by the addition of a mixture containing ammonia (55%) and deionised water (45%), and ultrasonication for 1 h. The resulting particles were added into the alcohol solution of $Ti[OCH(CH_3)_2]_4$, and the mixture was stirred for 30 min. After that, 0.4 mL HNO_3 was added and the mixture was vigorously stirred for 4 h and then ultrasonicated for 1 h. The solid was obtained by magnetic separation and thermal treated under nitrogen flow at 450 °C for 1 h.

After the synthesis, the obtained MNP were coated with carbon by CVD using the approach reported elsewhere [13]. Briefly, the MNP were thermal treated under nitrogen flow up to 400 °C and then the reduction step was carried out in a hydrogen atmosphere for 2 h. After that, the temperature was increased up to 750 °C for sample C@MNP_CVD750, or 850 °C for sample C@MNP_CVD850 (where MNP can be FeO, CoFeO, CuFeO or TiFeO) to coat the magnetic nanoparticles with carbon using ethane as a carbon precursor.

The samples were characterised by different techniques. The BET surface areas of the prepared samples were calculated from N_2 equilibrium adsorption/desorption isotherms, determined to be −196 °C with a Quantachrome Instruments NOVA 4200e apparatus. TG analysis was carried out to determine the relative amount of metals in the samples, and for that purpose, an STA 409 PC/4/H Luxx Netzsch thermal analyser was used. TEM micrographs were obtained using an LEO 906E microscope operating with an accelerating voltage of 120 kV. The phase purity of the samples was inferred by XRD analyses, over a range of $2\theta = 20°–80°$ with a Philips X'Pert MPD diffractometer (Cu-Ka = 0.15406 nm). The Rietveld refinement was used to evaluate the results.

3.2. Photocatalytic Ozonation Reactions

Experiments of PCO were carried out in a glass immersion photochemical reactor loaded with 0.5 g L^{-1} of catalyst concentration, as described elsewhere [30]. The reactions were performed with the following initial concentrations: $C_{0,\ OMA}$ = 89.05 mg L^{-1}, $C_{0,\ ANL}$ = 93.13 mg L^{-1}, $C_{0,\ MTLC}$ = 283.80 g L^{-1} and $C_{0,\ DYE}$ = 100 mg L^{-1}. The radiation source was a Heraeus TQ 150 medium-pressure mercury vapour lamp placed axially with a DURAN 50® glass cooling jacket positioned around the lamp, resulting in the main emission lines at λ_{exc} = 365 nm, 405 nm, 436 nm, 546 nm and 578 nm, which correspond to a spectral energy range between 2.15 eV and 3.40 eV. The gas was bubbled into the reactor at a constant flow rate (150 cm^3 min^{-1}) and inlet O_3 concentration (50 g m^{-3}). O_3 was produced from pure O_2 in a BMT 802X ozone generator. The concentration of O_3 in the gas phase was monitored with a BMT 964 O_3 analyser. Due to the magnetic behaviour of the prepared samples, the catalysts were easily recovered from the solution through the application of a magnetic field after each reaction.

The performance of the different catalysts during OMA removal by PCO was evaluated in terms of pollutant concentration as determined by high performance liquid chromatography (HPLC) analysis with an Hitachi Elite Lachrom system equipped with an ultraviolet detector. The stationary phase was an Altech AO-100 column working at room temperature under isocratic elution with H_2SO_4 (5 mM) at a flow rate of 0.5 cm^3 min^{-1}. The injection volume was 15 μL, and the detector wavelength was 200 nm [30]. The concentration of DYE was determined bya PG Instruments T60 UV-Vis spectrophotometer at the maximum absorption wavelength (630 nm). ANL and MTLC concentrations were determined using an HPLC Hitachi Elite Lachrom system equipped with a diode array detector (DAD). For ANL concentration, the stationary phase was a Lichrocart Purospher Star column working at room temperature under isocratic elution with 40% water and 60% methanol at a flow rate of 1 mL min^{-1}, an injection volume of 40 μL and the wavelength used was 200 nm. In the case of MTLC, the separation of the pollutant was attained by using the same column as used for ANL but with an isocratic mobile phase containing 60% acetonitrile and 40% water.

In the case of DYE, ANL and MTLC, the catalytic activity was also evaluated in terms of mineralization degree by the determination of TOC removal by a Shimadzu TOC-L total organic carbon analyser coupled with an ASI-L autosampler using the non-purgeable organic carbon (NPOC) method.

4. Conclusions

The use of magnetic nanoparticles (MNP) of FeO and their Ag and Cu bimetallic derivatives was successfully demonstrated for the photocatalytic ozonation (PCO) of oxamic acid (OMA), a model compound, for the first time. MNP also presented high catalytic activity during degradation of other representative pollutants.

The CVD coating procedure proves to be a simple and accessible technique to efficiently carbon-coat FeO MNP, as well as their bimetallic derivatives. This procedure allowed us to take advantage of the catalytic properties of the carbon materials, in addition to the efficient recovery attained due to their magnetic character.

The increase in performance associated with the temperature of CVD demonstrated the importance of the carbon coating in the photocatalytic process. Notably, it showed that more critical than the magnetic core was the quality (amount) of the carbon coating. The best results correlate with the increasing carbon content of the surface associated with higher CVD temperatures.

The modification of the FeO core with other metals did not represent an advantage for the carbon-coated MNP. The pristine C@FeO_CVD850 displayed an obvious synergy in PCO owing to the thermal and coating effects. C@FeO_CVD850 was quickly removed from the media and successfully applied in consecutive runs, presenting a rapid recycling operation mode due to its magnetic character. The successive PCO experimental runs with OMA showed that the C@FeO_CVD850 photocatalyst is very stable, remaining highly active after three initial cyclic tests.

The PCO of other pollutants (model compound of a textile effluent, an industrial pollutant and an emergent contaminant) revealed that the synergic effect evidenced by C@FeO_CVD850 enhanced the organic matter removal of the aqueous solutions compared to single ozonation, with almost complete mineralization being achieved.

Compared to the photocatalytic performance of commercial P25 (TiO2), no significant differences were observed, suggesting that C@FeO_CVD850 emerged as a remarkable alternative to remove a wide range of pollutants by PCO, due to its additional capacity of being efficiently recovered from the solution (by means of its magnetic properties), which is very appealing in terms of practical applications. Therefore, this work successfully answered to the constant challenge of developing a catalyst with similar performance to P25.

In conclusion, C@FeO_CVD850 emerges as a promising catalyst in the PCO process of an extensive class of pollutants with the advantage of being easily recovered from the media.

Author Contributions: Conceptualization, C.A.O. and O.S.G.P.S.; Investigation, C.A.O. and P.S.F.R.; Resources, M.F.R.P. and J.L.F.; Writing—Original Draft Preparation, C.A.O.; Validation and Writing—Review and Editing, O.S.G.P.S., M.F.R.P. and J.L.F.

Funding: This work is a result of: Project "AIProcMat@N2020-Advanced Industrial Processes and Materials for a Sustainable Northern Region of Portugal 2020", with the reference NORTE-01-0145-FEDER-000006, supported by Norte Portugal Regional Operational Programme (NORTE 2020), under the Portugal 2020 Partnership Agreement, through the European Regional Development Fund (ERDF); Associate Laboratory LSRE-LCM-UID/EQU/50020/2019-funded by national funds through FCT/MCTES (PIDDAC).

Conflicts of Interest: The authors declare no conflict of interest.

References

1. Mehrjouei, M.; Müller, S.; Möller, D. A review on photocatalytic ozonation used for the treatment of water and wastewater. *Chem. Eng. J.* **2015**, *263*, 209–219. [CrossRef]
2. Chávez, A.M.; Ribeiro, A.R.; Moreira, N.F.F.; Silva, A.M.T.; Rey, A.; Álvarez, P.M.; Beltrán, F.J. Removal of Organic Micropollutants from a Municipal Wastewater Secondary Effluent by UVA-LED Photocatalytic Ozonation. *Catalysts* **2019**, *9*, 472. [CrossRef]
3. Mills, A.; Lee, S.K. A web-based overview of semiconductor photochemistry-based current commercial applications. *J. Photochem. Photobiol. A Chem.* **2002**, *152*, 233–247. [CrossRef]
4. Chen, X.; Gambhir, S.S.; Cheon, J. Theranostic Nanomedicine. *Acc. Chem. Res.* **2011**, *44*, 841. [CrossRef] [PubMed]
5. Xie, J.; Liu, G.; Eden, H.S.; Ai, H.; Chen, X. Surface-Engineered Magnetic Nanoparticle Platforms for Cancer Imaging and Therapy. *Acc. Chem. Res.* **2011**, *44*, 883–892. [CrossRef]
6. Suber, L.; Marchegiani, G.; Olivetti, E.; Celegato, F.; Coisson, M.; Tiberto, P.M.; Allia, P.; Barrera, G.; Pilloni, L.; Barba, L.; et al. Pure magnetic hard fct FePt nanoparticles: Chemical synthesis, structural and magnetic properties correlations. *Mater. Chem. Phys.* **2014**, *144*, 186–193. [CrossRef]
7. Long, Y.; Xie, M.; Niu, J.; Wang, P.; Ma, J. Preparation of acid–base bifunctional core–shell structured Fe3O4@SiO2 nanoparticles and their cooperative catalytic activity. *Appl. Surf. Sci.* **2013**, *277*, 288–292. [CrossRef]
8. Xu, X.; Wu, H.; Li, Z.; Sun, X.; Wang, Z. Iron oxide-silver magnetic nanoparticles as simple heterogeneous catalysts for the direct inter/intramolecular nucleophilic substitution of π-activated alcohols with electron-deficient amines. *Tetrahedron* **2015**, *71*, 5254–5259. [CrossRef]
9. Shokouhimehr, M. Magnetically Separable and Sustainable Nanostructured Catalysts for Heterogeneous Reduction of Nitroaromatics. *Catalyst* **2015**, *5*, 534–560. [CrossRef]
10. Hu, J.; Chen, G.; Lo, I.M.C. Selective Removal of Heavy Metals from Industrial Wastewater Using Maghemite Nanoparticle: Performance and Mechanisms. *J. Environ. Eng.* **2006**, *132*, 709–715. [CrossRef]
11. Ali, Q.; Ahmed, W.; Lal, S.; Sen, T. Novel Multifunctional Carbon Nanotube Containing Silver and Iron Oxide Nanoparticles for Antimicrobial Applications in Water Treatment. *Mater. Today Proc.* **2017**, *4*, 57–64. [CrossRef]
12. Funari, V.; Mantovani, L.; Vigliotti, L.; Tribaudino, M.; Dinelli, E.; Braga, R. Superparamagnetic iron oxides nanoparticles from municipal solid waste incinerators. *Sci. Total Environ.* **2018**, *621*, 687–696. [CrossRef]
13. Pereira, L.; Dias, P.; Soares, O.; Ramalho, P.; Pereira, M.; Alves, M.; Pereira, M.F. Synthesis, characterization and application of magnetic carbon materials as electron shuttles for the biological and chemical reduction of the azo dye Acid Orange 10. *Appl. Catal. B Environ.* **2017**, *212*, 175–184. [CrossRef]
14. Mahmoodi, N.M. Photocatalytic ozonation of dyes using copper ferrite nanoparticle prepared by co-precipitation method. *Desalination* **2011**, *279*, 332–337. [CrossRef]
15. Mahmoodi, N.M.; Bashiri, M.; Moeen, S.J. Synthesis of nickel–zinc ferrite magnetic nanoparticle and dye degradation using photocatalytic ozonation. *Mater. Res. Bull.* **2012**, *47*, 4403–4408. [CrossRef]
16. Yin, J.; Liao, G.; Zhou, J.; Huang, C.; Ling, Y.; Lu, P.; Li, L. High performance of magnetic $BiFeO_3$ nanoparticle-mediated photocatalytic ozonation for wastewater decontamination. *Sep. Purif. Technol.* **2016**, *168*, 134–140. [CrossRef]
17. Figueiredo, J.L.; Pereira, M.F.R. Carbon as Catalyst. In *Carbon Materials for Catalysis*; Serp, P., Figueiredo, J.L., Eds.; John Wiley & Sons, Inc.: Hoboken, NJ, USA, 2009; pp. 177–217.
18. Cai, Y.; Qin, J.; Wei, F.; Xu, C.; Yao, Y.; Lu, F.; Wang, S. Magnetic $ZnFe_2O_4$–C_3N_4 Hybrid for Photocatalytic Degradation of Aqueous Organic Pollutants by Visible Light. *Ind. Eng. Chem. Res.* **2014**, *53*, 17294–17302.
19. Hussain, I.; Zhang, Y.; Li, M.; Huang, S.; Hayat, W.; He, L.; Du, X.; Liu, G.; Du, M. Heterogeneously degradation of aniline in aqueous solution using persulfate catalyzed by magnetic $BiFeO_3$ nanoparticles. *Catal. Today* **2018**, *310*, 130–140. [CrossRef]
20. De Oliveira, J.S.; Salla, J.D.; Kuhn, R.C.; Jahn, S.L.; Foletto, E.L. Catalytic Ozonation of Melanoidin in Aqueous Solution over $CoFe_2O_4$ Catalyst. *Mater. Res. Ibero Am. J.* **2019**, *22*, 1–9. [CrossRef]

21. Wang, J.; Shao, X.; Zhang, Q.; Ma, J.; Ge, H. Preparation and photocatalytic application of magnetic Fe2O3/SBA-15 nanomaterials. *J. Mol. Liq.* **2018**, *260*, 304–312. [CrossRef]
22. Wang, Z.; Ma, H.; Zhang, C.; Feng, J.; Pu, S.; Ren, Y.; Wang, Y. Enhanced catalytic ozonation treatment of dibutyl phthalate enabled by porous magnetic Ag-doped ferrospinel $MnFe_2O_4$ materials: Performance and mechanism. *Chem. Eng. J.* **2018**, *354*, 42–52. [CrossRef]
23. Ortiz-Quiñonez, J.L.; Pal, U.; Villanueva, M.S. Structural, Magnetic, and Catalytic Evaluation of Spinel Co, Ni, and Co–Ni Ferrite Nanoparticles Fabricated by Low-Temperature Solution Combustion Process. *ACS Omega* **2018**, *3*, 14986–15001. [CrossRef]
24. Vinosel, V.M.; Anand, S.; Janifer, M.A.; Pauline, S.; Dhanavel, S.; Praveena, P.; Stephen, A. Enhanced photocatalytic activity of Fe3O4/SnO2 magnetic nanocomposite for the degradation of organic dye. *J. Mater. Sci. Mater. Electron.* **2019**, *30*, 9663–9677. [CrossRef]
25. Shi, J.; Feng, S.; Chen, T.; Wu, F.; Guo, W.; Li, Y.; Zhao, P. Novel rattle-type magnetic Fe_3O_4@Ag@H-BiOCl photocatalyst with enhanced visible light-driven photocatalytic activity. *J. Mater. Sci. Mater. Electron.* **2018**, *29*, 10204–10213. [CrossRef]
26. Orge, C.; Faria, J.; Pereira, M.; Pereira, M.F. Removal of oxalic acid, oxamic acid and aniline by a combined photolysis and ozonation process. *Environ. Technol.* **2014**, *36*, 1075–1083. [CrossRef]
27. Sampaio, M.J.; Lima, M.J.; Baptista, D.L.; Silva, A.M.; Silva, C.G.; Faria, J.L. Ag-loaded ZnO materials for photocatalytic water treatment. *Chem. Eng. J.* **2017**, *318*, 95–102. [CrossRef]
28. Yao, Y.; Li, G.; Ciston, S.; Lueptow, R.M.; Gray, K.A. Photoreactive TiO_2/carbon nanotube composites: Synthesis and reactivity. *Environ. Sci. Technol.* **2008**, *42*, 4952–4957. [CrossRef]
29. Kumar, M.; Ando, Y. Chemical vapor deposition of carbon nanotubes: A review on growth mechanism and mass production. *J. Nanosci. Nanotechnol.* **2010**, *10*, 3739–3758. [CrossRef]
30. Orge, C.; Pereira, M.F.; Faria, J. Photocatalytic ozonation of model aqueous solutions of oxalic and oxamic acids. *Appl. Catal. B Environ.* **2015**, *174*, 113–119. [CrossRef]
31. Orge, C.A.; Soares, O.S.G.; Faria, J.L.; Pereira, M.F.R. Synthesis of TiO_2-Carbon Nanotubes through ball-milling method for mineralization of oxamic acid (OMA) by photocatalytic ozonation. *J. Environ. Chem. Eng.* **2017**, *5*, 5599–5607. [CrossRef]
32. Orge, C.; Faria, J.; Pereira, M.; Pereira, M.F.; Orge, C. Photocatalytic ozonation of aniline with TiO_2-carbon composite materials. *J. Environ. Manag.* **2017**, *195*, 208–215. [CrossRef] [PubMed]
33. Orge, C.; Pereira, M.; Faria, J.; Orge, C.; Pereira, M.F. Photocatalytic-assisted ozone degradation of metolachlor aqueous solution. *Chem. Eng. J.* **2017**, *318*, 247–253. [CrossRef]
34. Tang, Y.; Zhang, G.; Liu, C.; Luo, S.; Xu, X.; Chen, L.; Wang, B. Magnetic TiO_2-graphene composite as a high-performance and recyclable platform for efficient photocatalytic removal of herbicides from water. *J. Hazard. Mater.* **2013**, *252*, 115–122. [CrossRef] [PubMed]

Article

Heterogeneous Fenton-Like Degradation of *p*-Nitrophenol over Tailored Carbon-Based Materials

O. S. G. P. Soares [1,*], Carmen S. D. Rodrigues [2,*], Luis M. Madeira [2] and M. F. R. Pereira [1]

1 LSRE-LCM—Laboratory of Separation and Reaction Engineering—Laboratory of Catalysis and Materials, Faculty of Engineering, University of Porto, Rua Dr. Roberto Frias, 4200-465 Porto, Portugal; fpereira@fe.up.pt

2 LEPABE—Laboratory for Process Engineering, Environment, Biotechnology and Energy, Faculty of Engineering, University of Porto, Rua Dr. Roberto Frias, 4200-465 Porto, Portugal; mmadeira@fe.up.pt

* Correspondence: salome.soares@fe.up.pt (O.S.G.P.S.); csdr@fe.up.pt (C.S.D.R.); Tel.: +351-220-414-874 (O.S.G.P.S.); +351-220-414-851 (C.S.D.R.)

Received: 28 January 2019; Accepted: 9 March 2019; Published: 14 March 2019

Abstract: Activated carbon (AC), carbon xerogel (XG), and carbon nanotubes (CNT), with and without N-functionalities, were prepared. Catalysts were obtained after impregnation of these materials with 2 wt.% of iron. The materials were characterized in terms of N_2 adsorption at −196 °C, elemental analysis (EA), and the pH at the point of zero charge (pH_{PZC}). The *p*-nitrophenol (PNP) degradation and mineralization (assessed in terms of total organic carbon–TOC–removal) were evaluated during adsorption, catalytic wet peroxidation (CWPO), and Fenton process. The textural and chemical properties of the carbon-based materials play an important role in such processes, as it was found that the support with the highest surface area -AC- presents the best performance in adsorption, whereas the materials with the highest mesopore surface area -XG or Fe/XG- lead to best removals by oxidation processes (for XG it was achieved 39.7 and 35.0% and for Fe/XG 45.4 and 41.7% for PNP and TOC, respectively). The presence of N-functionalities increases such removals. The materials were reused in consecutive cycles: the carbon-based materials were deactivated by hydrogen peroxide, while the catalysts showed high stability and no Fe leaching. For the support with superior performances -XG-, the effect of nitrogen content was also evaluated. The removals increase with the increase of the nitrogen content, the maximum removals (81% and 65% for PNP and TOC, respectively) being reached when iron supported on a carbon xerogel doped with melamine was used as catalyst.

Keywords: carbon xerogel; carbon nanotubes; activated carbon; adsorption; catalytic wet peroxidation; heterogeneous Fenton's oxidation; *p*-nitrophenol

1. Introduction

Nowadays, the research and development of efficient wastewater treatment technologies have received huge attention from the scientific community, especially the treatment methods called advanced oxidation processes (AOPs) [1,2]. Among the innumerous AOPs, the Fenton reaction stands out, which has been widely studied in the treatment of effluents because it is recognized as a "green technology" (i.e., this process uses environmentally friendly reagents), implies low capital cost, is able to operate at atmospheric pressure and room temperature with a simple technology, is non selective (i.e., degrade various pollutants), improves the biodegradability and reduces the toxicity of the effluents, and is an efficient process [3–5].

The high efficiency of Fenton reaction in degradation of organic matter is due to the utilization of hydroxyl radicals, which are the second species with highest oxidation potential (2.8 eV) [6]. These radicals result from the catalytic decomposition of hydrogen peroxide, at low pH values, in the presence of iron (II) species (Equation (1)) [7]:

$$Fe^{2+} + H_2O_2 \rightarrow Fe^{3+} + HO^{\bullet} + OH^- \quad (1)$$

However, other reactions also occur at same time, for instance: (i) iron (III) reacts with hydrogen peroxide (Equation (2)) or $HO_2^{\bullet}$ (Equation (3)) and forms iron (II) again—catalyst regeneration, and (ii) undesirable hydroxyl radicals scavenging by excess of hydrogen peroxide (Equation (4)) and/or iron (II) (Equation (5)) [7]:

$$Fe^{3+} + H_2O_2 \rightarrow Fe^{2+} + HO_2^{\bullet} + H^+ \quad (2)$$

$$Fe^{3+} + HO_2^{\bullet} \rightarrow Fe^{2+} + O_2 + H^+ \quad (3)$$

$$HO^{\bullet} + H_2O_2 \rightarrow H_2O + HO_2^{\bullet} \quad (4)$$

$$HO^{\bullet} + Fe^{2+} \rightarrow Fe^{3+} + HO^- \quad (5)$$

The principal disadvantage of the homogeneous Fenton process is the introduction of iron into the treated water, so an additional separation step is required to remove/recover it. An alternative strategy, in order to overcome this drawback, is the immobilization of the catalyst on a porous solid matrix. Distinct materials, like zeolites, pillared clays, silica, silicalites, and carbons have been used to support iron [8–14].

Carbon materials are widely used as supports because they are very flexible materials, since their textural and chemical properties can be easily tailored by physical or chemical treatments [15,16]. Among the carbon materials, activated carbon (AC) is the most reported in the literature as iron support to be used as a catalyst for the Fenton reaction [13,14,17–20]. Recently, carbon nanotubes (CNT) [21–23] and carbon xerogels (XG) [24] have received intensive attention as iron supports due to their interesting properties when compared with other carbon-based materials, such as mechanical resistance, low limitations to mass transfer, high thermal stability in oxidation conditions, high electronic properties for CNT [25], high purity, mesoporous structure with good pore distribution, and high surface area for XG [26].

The presence of nitrogen groups on the surface of carbon-based materials improves their catalytic performance for advanced oxidation processes. This is explained by the increase of the electronic density, and consequently the surface basicity, which favors the interaction between the organic compounds and the support [17].

p-Nitrophenol (PNP) was used in this work as a model compound because it is commonly present in several industrial wastewaters such as those resulting from manufacturing of drugs, dyes, fungicides, and explosives [27]. PNP is toxic, carcinogenic, and mutagenic, needing to be removed from effluents before their discharge into water bodies. Due to its toxic nature to microorganisms, biological degradation is not suitable or presents slow degradation rates [28,29]; so, the use of AOPs, particularly of heterogeneous Fenton-like process, is of vital importance. There are some studies where this process is applied to degrade PNP using as catalysts acid-activated fly ash [30], activated carbon impregnated with iron and doped with nitrogen [17], CuO/zeolite [31], transition metals (such as Ti, Cr, Mn, Co and Ni) on magnetite [32], magnetite/silica microspheres [33], and magnetic Fe^0/Fe_3O_4/coke composite [34].

The main aim of this work is to evaluate the effect of the carbon-based materials nature (AC, CNT, and XG) with different textural properties and nitrogen functionalities in adsorption, catalytic wet peroxidation (CWPO) and heterogeneous Fenton oxidation of PNP. To the best of our knowledge, there are no studies combining the evaluation of all these parameters under the same experimental conditions.

2. Results and Discussion

2.1. Carbon-Based Materials Characterization

The results shown in Table 1 clearly show that the activated carbon sample presents the largest BET surface area - S_{BET} - (824 m^2 g^{-1}), while the carbon nanotubes are the carbon-based materials with the lowest surface area (273 m^2 g^{-1}). In addition, activated carbons are mainly microporous materials, whereas carbon nanotubes have a mesoporous nature and carbon xerogels present intermediate textural properties with a BET surface area of 572 m^2 g^{-1} and a mesoporous area (S_{meso}) of 290 m^2 g^{-1}. The N-doping with melamine or urea induces a reduction of the surface area, suggesting that the N-groups introduced may block the access of N_2 molecules to the pores. There is only a slight decrease of the surface area of the iron catalysts (Fe/y, where y is the carbon material used as support) compared to the unloaded carbon-based materials, confirming that the textural properties of the supported catalysts are not significantly different from the corresponding supports. Also, in this case, access to the pores may be partially blocked by iron species.

Table 1. Textural properties of the carbon-based materials *.

Sample	S_{BET} ** ($\pm$10 m^2 g^{-1})	S_{meso} ** ($\pm$10 m^2 g^{-1})	V_{micro} ** ($\pm$0.01 cm^3 g^{-1})	$V_{p\ P/Po=0.95}$ ** (cm^3 g^{-1})
AC	824	196	0.286	0.492
ACM	730	174	0.229	0.408
CNT	273	273	0	0.573
CNTM	222	222	0	0.512
XG	572	290	0.150	0.497
XGM	510	117	0.182	0.398
XGU	513	174	0.158	0.486
Fe/AC	807	179	0.288	0.479
Fe/ACM	717	171	0.251	0.408
Fe/CNT	266	266	0	0.540
Fe/CNTM	205	205	0	0.471
Fe/XG	564	273	0.149	0.491
Fe/XGM	490	133	0.168	0.402
Fe/XGU	486	166	0.147	0.449

* AC—activated carbon; CNT—carbon nanotubes; XG—xerogel; M—stands for melamine-doped materials; U—stands for urea-doped materials; Fe—stands for iron-containing catalysts. ** S_{BET}—surface area; S_{meso}—mesopore surface area; V_{micro}—micropore volume, V_p—total specific pore volume.

The doping with melamine as N-precursor results in the incorporation of a large amount of nitrogen (between 3.4% and 5.3%) on the surface of the carbon-based materials, as determined by elemental analysis (N_{EA}—Table 2). In the case of the carbon xerogels, for which two types of nitrogen precursors were used, it was observed that the doping with melamine allows the incorporation of more nitrogen (4.51%) than urea (3.30%), since the molecule of melamine has more nitrogen atoms (six) than urea (two). The N1s spectrum, obtained by X-ray photoelectron spectroscopy (XPS), revealed the presence of quaternary nitrogen (N-Q), pyrrole (N-5), and pyridinic (N-6) structures on samples ACM and CNTM, whereas samples XGU and XGM only present N-5 and N-6 groups. The N-6 group (B.E. = 399.0 $\pm$ 0.2 eV) is the most abundant species in all samples, followed by the N-5 group (B.E. = 400.4 $\pm$ 0.3 eV). The corresponding binding energies and percentages of the N-functionalities are presented in Table 2. All the carbon-based materials present a neutral/basic character, with the pH at the point of zero charge (pH_{pzc}) values between 7.0 and 8.8.

Table 2. Chemical properties of the carbon-based materials *.

Sample	N Content		N-6 **		N-5 **		N-Q **		pH_{PZC} (±0.1)
	N_{EA} (wt.%)	N_{XPS} (wt.%)	B.E. (eV)	(wt.%)	B.E. (eV)	(wt.%)	B.E. (eV)	(wt.%)	
AC	0.15	-	-	-	-	-	-	-	7.5
ACM	3.40	3.14	398.4	1.68	400.0	1.07	401.4	0.39	7.5
CNT	0.01	-	-	-	-	-	-	-	7.0
CNTM	5.31	4.8	398.9	2.24	400.3	1.66	401.5	0.87	7.5
XG	0	-	-	-	-	-	-	-	8.0
XGM	4.51	4.64	398.5	2.48	400.6	2.16	-	-	8.8
XGU	3.30	3.87	398.6	2.18	400.7	1.70	-	-	8.3

* AC—activated carbon; CNT—carbon nanotubes; XG—xerogel; M—stands for melamine-doped materials; U—stands for urea-doped materials. ** N-6—pyridinic nitrogen structure; N-5—pyrrole nitrogen structure; N-Q—quaternary nitrogen structure.

2.2. Adsorption in Carbon-Based Materials

First, the adsorption of PNP in the supports (AC, CNT, and XG) and supports doped with melamine (ACM, CNTM, and XGM) was assessed. In Figure 1a,b it is possible to observe that PNP and TOC removals increase in the following order: CNT < XG < AC materials. Ribeiro et al. [35] observed the same tendency for adsorption of 2-nitrophenol. The amount of PNP adsorbed increases when N-doped materials are used, maintaining the same trend. For all samples PNP and TOC removals are nearly the same, evidencing that only adsorption phenomenon is occurring.

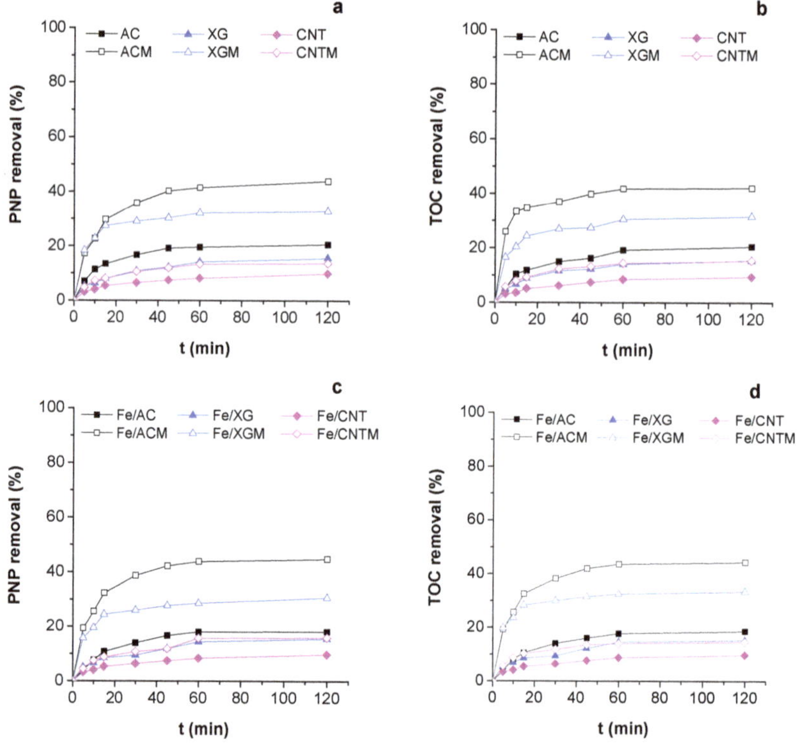

Figure 1. p-nitrophenol (PNP) and total organic (TOC) removal during adsorption with the carbon supports (**a**,**b**) and with the iron-containing catalysts (**c**,**d**) (pH = 3.0, T = 30 °C, [support] = 0.25 g L^{-1} and [PNP] = 3.6 mmol L^{-1}).

The adsorption capacity of the materials is related to their specific surface area (if surface chemistry properties of the carbon materials are similar), as evidenced by the fact that the lower amount of PNP adsorbed was obtained when CNT was used, which presents the lowest BET surface area (273 $m^2 g^{-1}$—see Table 1), while the highest amount was obtained with the AC support that presents the highest surface area (824 $m^2 g^{-1}$—see Table 1); the same tendency was reported by Messele et al. [36], who found that the highest adsorption capacity towards phenol was obtained with the carbon xerogel with the largest surface area. The adsorption results confirm that the textural properties of the supports play an important role in adsorption of PNP, this being demonstrated by the strong linear correlation found between PNP removal after 2 h and the materials BET surface areas (determination coefficient (r^2) > 0.999; Figure S1a—see Supporting Information section). However, other factors also play an important role, namely the surface chemistry.

The presence of N-containing groups on the surface of the supports led to an increase in their adsorption capacity, since the removal of PNP and TOC increased—see Figure 1a,b. This improvement is related to the N-functionalities incorporated on the surface of the materials, which allow to increase the surface basicity of the supports, i.e., increase the π electronic density of the carbons layers [37], favoring the interaction between the aromatic compound and supports. Similar results were observed by other authors [17,20,36–39]. In view of these results, it can be concluded that the surface chemistry properties also play a relevant role in the adsorption process (see Figure S1b; r^2 > 0.99).

The same tendency was observed for adsorption runs with supports impregnated with iron. The PNP and TOC removals shown in Figure 1c,d, reached with iron-containing catalysts, were similar to those obtained with the supports. This result is in accordance with the expected trend, since the catalysts and corresponding supports present similar BET surface areas (see Table 1) and surface chemistries.

2.3. Catalytic Wet Peroxidation (CWPO) Using the Supports as Catalysts

In order to evaluate the catalytic performance of the carbon-based materials (without iron) in peroxidation and mineralization of PNP, experiments were carried out in the presence of each support and oxidizing agent (hydrogen peroxide). The PNP and TOC removals are shown in Figure 2.

The PNP and TOC removals were very low (11.0% and 8.5%, respectively, after 2 h—Figure 2a,b) in the presence of the oxidant alone due to the fact that H_2O_2 has a low oxidation potential (1.8 eV). As shown in Figure 2, the presence of AC, CNT, or XG supports increases the efficiency of the process comparatively to either adsorption (Figure 1) or experiments where only oxidant has been used (Figure 2), which is due to the formation of hydroxyl radicals (see Figure 3a) by the decomposition of hydrogen peroxide on the surface of the carbon-based materials. This is in line with a previous work [17] and in accordance with the radical mechanism of catalytic decomposition of H_2O_2 using activated carbon (Equation (6)) proposed by Santos et al. [40]. The PNP and TOC removals increase in the following order: XG > AC > CNT. Nevertheless, it should be taken into account that in these experiments, part of the PNP (and organic compounds formed) are adsorbed on the carbon-based materials, so that there is the contribution of both reaction and adsorption; therefore, the PNP and TOC removals obtained by reaction only were estimated discounting from those data the removals reached by adsorption (i.e., without oxidant). Such data are shown in Figure 2c, being noticed that the removals reached by reaction are higher for the XG, followed by the CNT and finally the AC. XG and CNT present better catalytic performances probably due to their mesoporous nature; in fact, they present higher mesoporous areas (290 and 273 $m^2 g^{-1}$ for XG and CNT, respectively—see Table 1) than AC. In addition, Figure 3a also reveals that the mesoporosity of the materials benefits the decomposition of H_2O_2 into hydroxyl radicals. Actually, in addition to the fastest consumption of hydrogen peroxide, kinetics of hydroxyl radicals formation (assessed through the blank experiments) is also fastest for the mesoporous materials XG or CNT; this behavior can be related to the decrease of internal diffusional limitations with the increase of mesoporosity.

$$ACM + H_2O_2 \rightarrow ACM^+ + OH^- + HO^\bullet \quad (6)$$

Again, the N-functionalities on the surface of the carbon-based materials benefit the CWPO (see Figure 2a,b) increasing the hydrogen peroxide consumption and hydroxyl radicals formation in blank runs (see Figure 3a,c), following the same tendency of undoped supports. Sun et al. [41] obtained a greater phenol removal by sulfate radicals when using N-doped carbon nanotubes as catalysts. Rodrigues et al. [17] and Messele et al. [38] also observed the positive effect of the presence of N-groups on the surface of activated carbon in PNP and phenol removal, respectively, by wet peroxidation. Finally, Santos et al. [42] found higher wet oxidation efficiency of phenol when using carbon nanotubes doped with melamine as catalyst.

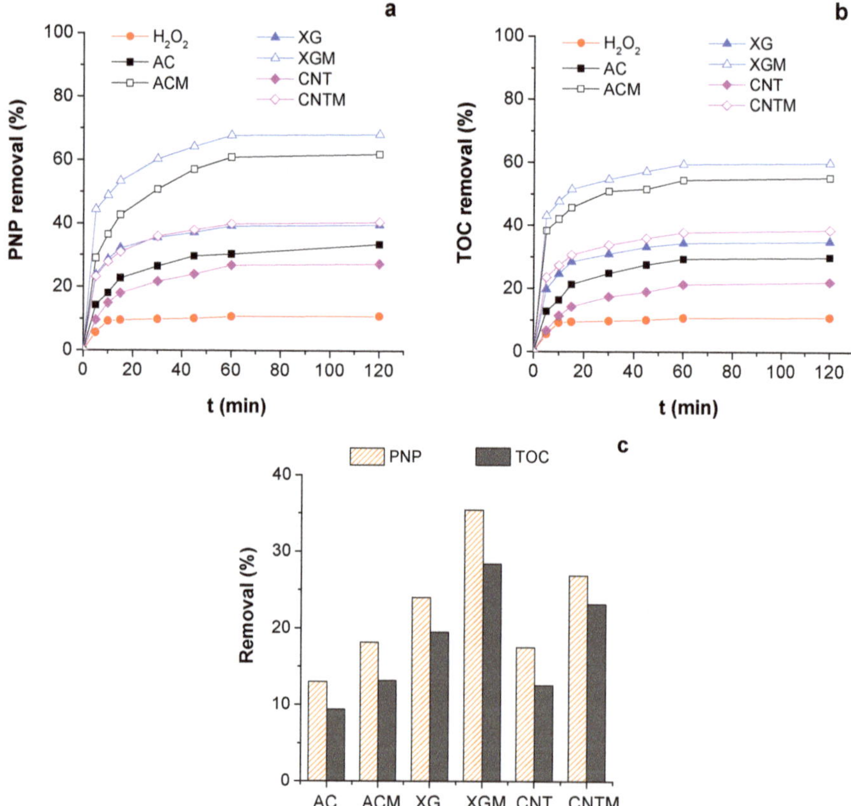

Figure 2. *p*-nitrophenol (PNP) (**a**) and total organic (TOC) (**b**) removal during catalytic wet peroxidation (CWPO) and adsorption, and PNP and TOC removals after 2 h for wet peroxidation only (**c**) using the carbon supports (pH = 3.0, T = 30 °C, [support] = 0.25 g L^{-1}, [H_2O_2] = 29 mmol L^{-1}, [PNP] = 3.6 mmol L^{-1} and H_2O_2:PNP molar ratio = 8.1).

For all the carbon supports, the presence of *p*-benzoquinone, hydroquinone, and *p*-nitrocatechol (higher molecular weight intermediate by-products reported for PNP oxidation [43–45]) was not observed during the wet peroxidation experiments. Thus, the difference between the removal of PNP and TOC is probably due to the formation of short-chain carboxylic acids (the last intermediate compounds formed), which are more difficult to oxidize, and therefore remain in solution contributing for TOC.

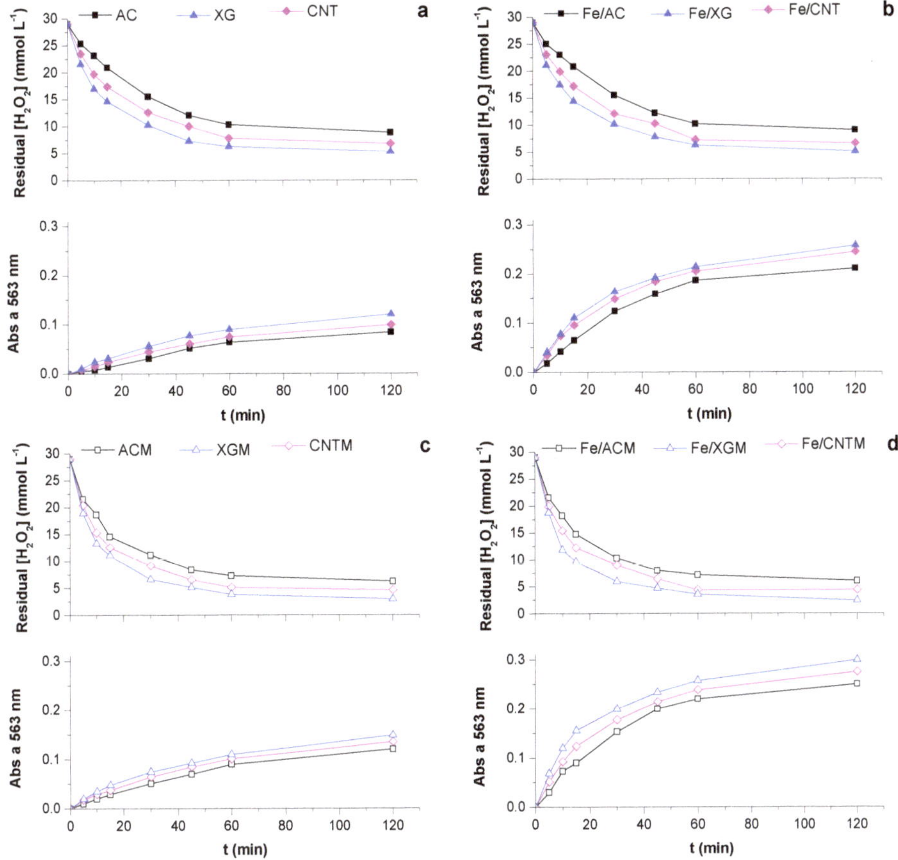

Figure 3. Consumption of hydrogen peroxide and hydroxyl radicals formation as a function of time for experiments with the supports (**a**) or with the iron supported catalysts (**b**) and supports (**c**) or catalysts (**d**) N-doped with melamine (pH = 3.0, T = 30 °C, [support or catalyst] = 0.25 g L^{-1}, $[H_2O_2]$ = 29 mmol L^{-1} and [PNP] = 0 mmol L^{-1}).

2.4. Fenton Reaction

The different supports were impregnated with 2.0 wt.% of iron (Fe/AC, Fe/CNT, Fe/XG, Fe/ACM, Fe/CNTM, and Fe/XGM) in order to assess its effect in PNP and TOC removal. Figure 4 reveals that both textural and chemical properties play an important role in the oxidative process. The evolution of PNP and TOC removals clearly indicates that high mesoporosity combined with high surface area benefit the process, which is in agreement with results shown in Figure 3b that evidence a high amount of hydroxyl radicals formed. The Fe/XG sample, which is the catalyst with the highest S_{meso} and intermediate S_{BET} (see Table 1), presents the best performance (41.7% and 45.4% for TOC and PNP removals, respectively), followed by the Fe/AC catalyst that has the larger S_{BET} but with a low S_{meso}, proceeded by the Fe/CNT catalyst which is the support with the lowest S_{BET}, although being a mesoporous material. Nevertheless, taking into account the fact that adsorption occurs simultaneously with Fenton reaction, this phenomenon was suppressed in order to assess the improvement achieved in the heterogeneous Fenton conditions. Figure 4c shows that a change in the trend of PNP and TOC removal improvements can be observed. Fe/XG catalyst remains the one with highest performance, followed by Fe/CNT and Fe/AC, this trend being similar to those observed in hydroxyl radicals formation and hydrogen peroxide consumption (see Figure 3b).

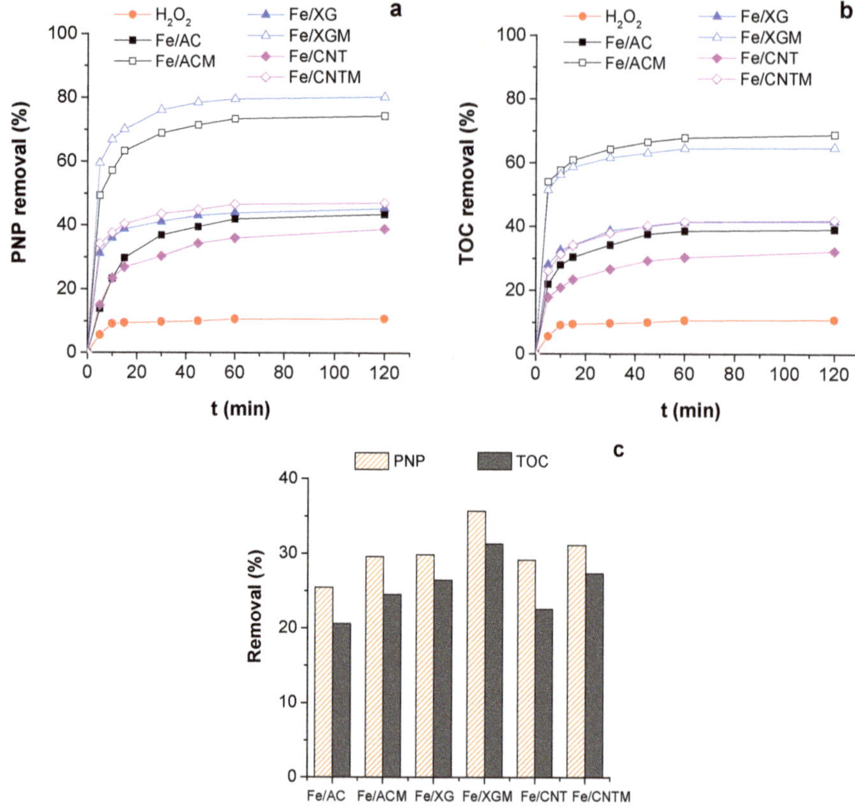

Figure 4. *p*-nitrophenol (PNP) (**a**) and total organic (TOC) (**b**) removal during Fenton's reaction and adsorption, and PNP and TOC removal, after 2 h, for Fenton reaction only (**c**) with the Fe-containing catalysts (pH = 3.0, T = 30 °C, [catalyst] = 0.25 g L^{-1}, [H_2O_2] = 29 mmol L^{-1}, [PNP] = 3.6 mmol L^{-1} and H_2O_2:PNP molar ratio = 8.1).

A considerable increase in the performances was observed when catalysts containing N-groups on the carbon surface were used, also highlighting the importance of the surface chemistry in the Fenton reaction. The same tendency was reported by Messele et al. [36] when evaluating the degradation of phenol by carbon xerogels doped with nitrogen and impregnated with zero-valent iron. The main reason for the process efficiency increase is related to the presence of N-functionalities on the carbon surface that increased the carbon basicity and, consequently, the content of electron rich sites on carbon basal planes, which increases the adsorption of the PNP besides promoting the easy access of the oxidant to the iron species that favors the hydroxyl radicals formation [38]. This fact was corroborated by the highest formation of hydroxyl radicals in blank runs with N-doped catalysts when compared with undoped catalysts (see Figure 3b,d). Figure 4 shows that the N-doped catalysts present the same trend as the undoped catalysts.

The PNP and TOC removals achieved by the Fenton process (in the ranges of ~43–80% and 32–69%, respectively) were slightly higher than the CWPO with the carbon supports (in the ranges of ~27–68% and ~22–60%, respectively). This is due to the fact that in the first process the formation of hydroxyl radicals occurs by: (i) decomposition of H_2O_2 on the carbon surface through Equation (6) [46] (which also occurs in the CWPO) and (ii) reaction between iron and oxidant (Equation (1) [7]). Experiments carried out in presence of a radical scavenging agent (dimethyl sulfoxide (DMSO) in a molar ratio of 1:10 to the oxidant) (see Figure S2 in the Supporting Information section) confirmed the occurrence of radical-mediated surface reactions because PNP and TOC removals obtained in these experiments were identical to those reached by simple adsorption. Other authors also reported that the presence of iron on carbon supports improves the degradation of organic compounds by oxidation [17,20,36,38,47].

Again, high molecular weight intermediate compounds (such as hydroquinone, *p*-benzoquinone, and *p*-nitrocatechol) were not detected in solution, so the differences between the PNP and TOC removals are probably due to the formation of refractory low molecular weight carboxylic acids during the reaction that remain in solution and contribute for TOC.

The results obtained show that both textural and chemical properties of the carbon-based materials had an important role in Fenton oxidation, where high mesoporosity combined with high surface area and N-functionalities improve their catalytic performance.

2.5. Supports and Catalysts Stability and Reusability

The stability and reusability of all the N-doped supports and Fe-containing catalysts were also evaluated. For that, three consecutive reaction runs were performed under the same operating conditions evaluating the PNP and TOC removals and iron leached in solution. At the end of each run, the supports and catalysts were recovered by filtration and reused in the next run after drying at 40 °C during 1 day.

Regarding the supports, it can be seen in Figure 5 that: (i) the performance of ACM decreases during the reutilization cycles, and (ii) for XGM and CNTM there is a great decrease in PNP and TOC removals from the first to the second cycle and a slight decrease to the third cycle. The decrease in efficiency of CWPO between reutilizations can be due to the porosity blocking and/or deactivation of carbons due to the oxidation of the carbon surface by the hydrogen peroxide, forming oxygen surface groups that have a negative influence in the electronic interactions between the carbon surface and PNP or other organic compounds. This deactivation of carbons was also reported in literature by other authors [25,48,49].

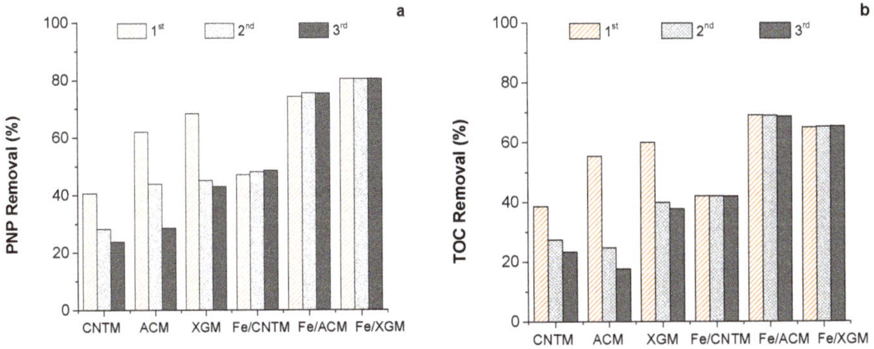

Figure 5. *p*-nitrophenol (PNP) (**a**) and total organic (TOC) (**b**) removal, after 2 h, in subsequent cycles during wet peroxidation and adsorption with the carbon supports or Fenton's reaction and adsorption with the iron-containing catalysts (pH = 3.0, T = 30 °C, [support] = 0.25 g L^{-1}, [H_2O_2] = 29 mmol L^{-1}, [PNP] = 3.6 mmol L^{-1} and H_2O_2:PNP molar ratio = 8.1).

On the other hand, no deactivation was observed when iron supported catalysts were used (see Figure 5), being the maximum variations between the three cycles, after 2 h of reaction, inferior to 1.4% and 0.3% for PNP and TOC removals, respectively. In addition, the iron concentration in solution was lower than the detection limit (0.28 mg/L) for all runs. Iron in solution was still not detected in additional experiments where a dose of catalyst 10 times higher was used, indicating that no iron leaching occurs (at least it is below the detection limit, which would correspond to 0.56 wt.%); consequently, the catalysts were stable, which is a crucial aspect for application at industrial scale.

2.6. Effect of Nitrogen Content

In order to evaluate the effect of the nitrogen content on the efficiency of the processes, new experiments were performed using a carbon xerogel doped with urea -XGU and Fe/XGU-, since carbon xerogel is shown to be the best support and urea allows the introduction of lower N-content than melamine. The carbon xerogel supports present similar textural properties but different N-content: XGM presents the highest N-amount among the N-doped supports, whereas XG does not contain nitrogen (see Table 2). XGM and XGU present the same type of N-groups, N-6 and N-5 structures. Figure 6a,b show that the XGU support presents an intermediate performance between XG and XGM during adsorption and CWPO, respectively; the same trend was observed when iron supported catalysts were applied in Fenton reaction (Figure 6c). This trend demonstrates that the N-content has an important role in the PNP and TOC removal, their performance being favored by the presence of higher amounts of N, which must be due to the increase of hydroxyl radicals formed. In fact, Figure 7 shows that the formation of hydroxyl radicals increases in the presence of samples with higher N-content, which is in line with other works [17,38].

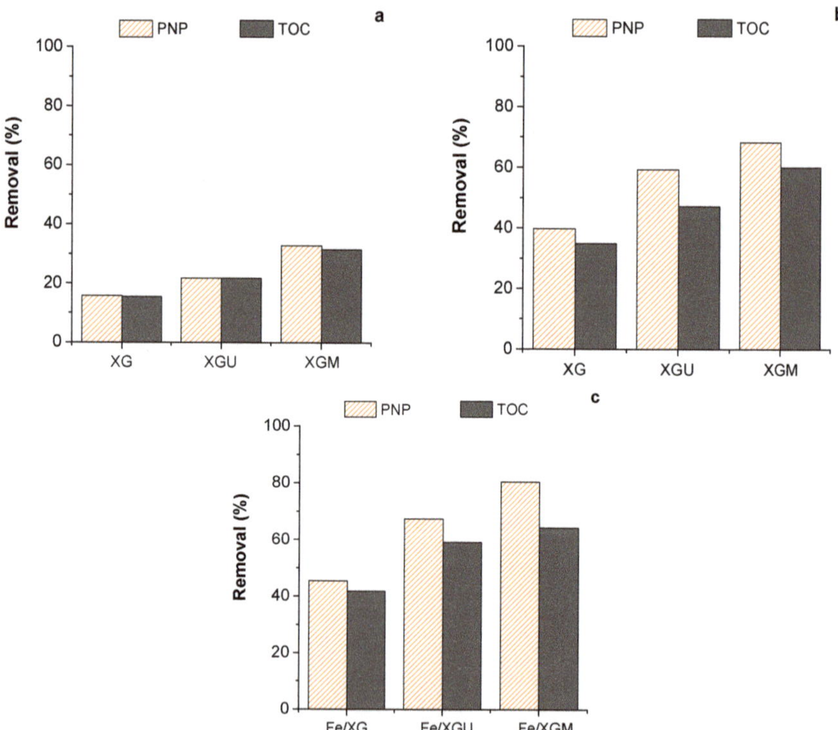

Figure 6. *p*-nitrophenol (PNP) and total organic (TOC) removal after 2 h of adsorption (**a**) and reaction (**b**) with support carbon xerogels (XG, XGU and XGM) and Fenton's reaction (**c**) using catalysts (Fe/XG, Fe/XGU and Fe/XGM) (pH = 3.0, T = 30 °C, [support or catalyst] = 0.25 g L^{-1}, $[H_2O_2]_{\text{when used}}$ = 29 mmol L^{-1}, [PNP] = 3.6 mmol L^{-1} and H_2O_2:PNP molar ratio = 8.1, when H_2O_2 was used).

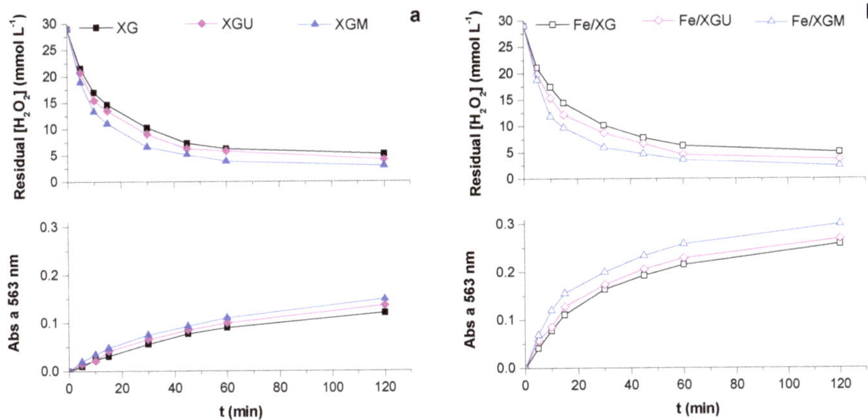

Figure 7. Consumption of hydrogen peroxide and hydroxyl radicals formation as a function of time for blank reaction with supports—xerogel (XG), XGU and XGM (**a**) and with catalysts -Fe/XG, Fe/XGU and Fe/XGM (**b**) (pH = 3.0, T = 30 °C, [support or catalyst] = 0.25 g L^{-1}, [H_2O_2] = 29 mmol L^{-1}, $[PNP]_{blank\ run}$ = 0 mmol L^{-1}).

A close to linear correlation between the PNP removal after 2 h with the N-content of XG, XGM, and XGU supports was found (determination coefficient (r^2) > 0.98; Figure S3—see Supplementary Materials), demonstrating the relationship between the nitrogen content and catalytic performance of the materials. The carbon xerogel with the highest N-content presents the highest catalytic activity.

2.7. Evaluation of Intermediates Compounds Formed during the Heterogeneous Fenton Reaction

The maximum mineralization achieved in the Fenton process was 16% lower than the PNP removal for the catalyst with the best catalytic performance -Fe/XGM- indicating that some intermediate compounds formed during the oxidative process remain in solution. Thus, a new run was carried out in the presence of this catalyst in order to assess the presence of these organic compounds and their contribution to TOC. Figure 8a shows that in the first 30 min of reaction the PNP concentration decreased quickly; then its degradation rate slowed down until 60 min, leaving its concentration practically constant afterwards; overall conversion was 80.3%. The mineralization achieved, after 120 min of reaction, was 64.7%—see Table 3. The removals achieved were very similar to those obtained previously.

Table 3. *p*-nitrophenol (PNP) and carboxylic acids concentration along the reaction time under optimized conditions with Fe/XGM as catalyst, total carbon based on such analyzed compounds, experimental total organic carbon (TOC) and carbon balance.

t (min)	PNP (mgC L^{-1})	Pyruvic (mgC L^{-1})	Maleic (mgC L^{-1})	Oxalic (mgC L^{-1})	Total Carbon * (mgC L^{-1})	TOC (mgC L^{-1})	Contribution ** (%)
0	260.7	0.0	0.0	0.0	260.7	260.9	99.9
5	105.3	7.3	6.4	5.1	124.1	126.3	98.3
10	86.3	9.0	8.6	6.3	110.1	112.4	98.0
15	78.0	11.0	10.9	7.5	107.5	107.3	100.2
30	62.1	13.9	12.8	11.0	99.9	100.8	99.1
45	55.6	13.0	12.0	13.9	94.5	94.6	99.9
60	52.7	12.5	11.8	13.1	90.1	92.8	97.1
120	50.9	12.4	11.7	13.0	87.4	92.2	95.5

* Total carbon assessed from the sum of the identified compounds; ** Ratio between Total carbon and TOC × 100.

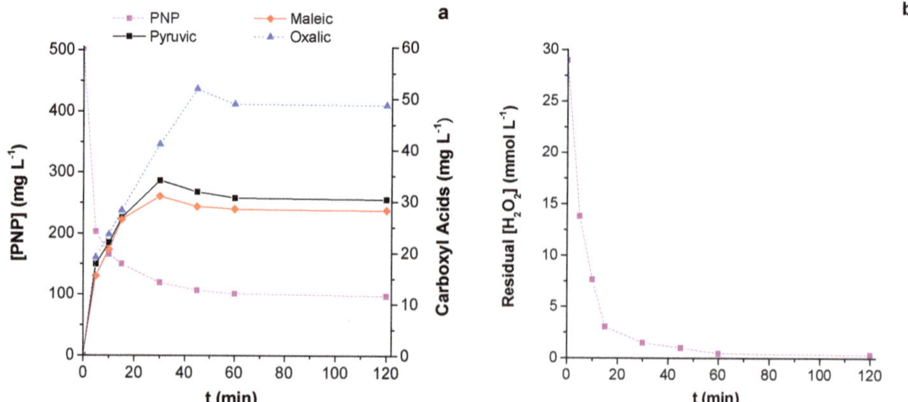

Figure 8. *p*-nitrophenol (PNP) and carboxylic acids concentration (**a**) and consumption of hydrogen peroxide (**b**) along Fenton reaction with Fe/XGM catalyst (pH = 3.0, T = 30 °C, [Fe/XGM] = 0.25 g L^{-1}, [H_2O_2] = 29 mmol L^{-1}, [PNP] = 3.6 mmol L^{-1} and H_2O_2:PNP molar ratio = 8.1).

Again, hydroquinone, *p*-benzoquinone, and *p*-nitrocatechol were not detected in the solution, indicating that their degradation into carboxylic acids is fast. This fact was corroborated by the carbon balance, which was higher than 95%—see Table 3. The absence of such compounds was also observed in another work where PNP has been degraded by the heterogeneous Fenton process catalyzed with activated carbon doped with melamine and impregnated with iron [17]. The presence of carboxylic acids in solution was observed and the concentrations of pyruvic, maleic, and oxalic acids is shown in Figure 8a. Their formation increased in the first 30 min of reaction, which corresponds to the time where a higher decrease of the PNP concentration occurs. Afterwards the concentration of pyruvic and maleic acids decreased due to their oxidation into oxalic acid, as proposed by other authors [17,44]. From 45 to 60 min of reaction, the concentration of oxalic decreased slightly due to its degradation into carbon dioxide and water. For reaction times higher than 60 min the concentration of carboxylic acids remains constant (corresponding also to the absence of PNP oxidation) indicating the cessation of the oxidation reactions; this is corroborated by the total consumption of the oxidant (see Figure 8b).

3. Materials and Methods

3.1. Materials

PNP was purchased from Fluka (reagent grade, CAS: 100-02-7). The oxidant (30% (w/v) H_2O_2 aqueous solution) used in this work was analytical reagent grade supplied by Fluka. The sulfuric acid (95–97%) and nitric acid (65%) were obtained from Panreac and the sodium hydroxide (99%) was purchased from Merck. Resorcinol (99%) and formaldehyde solution (37%) were purchased from Sigma-Aldrich. Melamine ($\geq$99%) and urea (99.5%) were obtained from Fluka and Acros Organics, respectively.

3.2. Carbon-Based Materials Preparation and Characterization

Nitrogen-free and nitrogen-doped carbon-based materials with different textural properties were used. For that, commercial NORIT GAC 1240 PLUS activated carbon, commercial Nanocyl NC 3100 multiwalled carbon nanotubes, as well as a carbon xerogel, were used as starting materials.

Initially, the activated carbon was ground with a mortar and sieved to a particle diameter lower than 0.1 mm. Then, sample AC and sample CNT were doped with nitrogen using melamine (M) as precursor. For that, 0.6 g of the original AC and CNT was dipped into 50 mL of a 1 mol L^{-1} aqueous solution of melamine, and stirred at room temperature during 24 h. Thereafter, the samples (ACM and

CNTM) were filtered, washed with distilled water and dried in an oven at 110 °C overnight. These samples were heat treated at 600 °C during 1 h under nitrogen flow.

The XG was prepared at pH 6.9 by the conventional sol-gel method, using the polycondensation of resorcinol and formaldehyde as described elsewhere [50]. Briefly, 25 g of resorcinol were mixed with 40 mL of distilled water under magnetic stirring. Then, the pH was adjusted to 6.9 and a few drops of 2 mol L^{-1} NaOH were added, followed by the addition of 34 mL of formaldehyde solution and the final pH was adjusted to 6.9 by addition of 0.1 mol L^{-1} HCl. The gelation was achieved in a water bath at 85 °C during 3 days. After this step, the gel was crushed and dried in an oven for 4 days: 1st day at 60 °C, 2nd day at 80 °C, 3rd day at 100 °C, and 4th day at 120 °C. The dried gel was carbonized under nitrogen flow (100 cm^3 STP min^{-1}) at constant heating rate of 2 °C min^{-1}, under the following conditions: (1) from room temperature to 200 °C and hold for 2 h, (2) up to 300 °C and hold for 1 h, (3) up to 500 °C and hold for 2 h, (4) up to 600 °C and hold for 2 h. Nitrogen-doped carbon xerogels were prepared using the protocol described elsewhere [51] adding melamine or urea (XGM and XGU, respectively) as nitrogen precursor during the synthesis. For that, 15 g of resorcinol and 3 g of melamine or 4 g of urea were mixed in 18 mL of distilled water and were heated up to 90 °C under stirring until melamine or urea was dissolved. After that, the solution was cooled until it reached room temperature and 20 mL of formaldehyde was added. The next steps were equal to those described to sample XG. All the carbon xerogel samples were ground to a particle size lower than 0.1 mm.

Iron catalysts supported in all the carbon-based materials (Fe/y, where y is the carbon material used as support) were prepared by the incipient wetness impregnation method, which was always conducted under ultrasonic mixing. The precursor solution was added drop wise using a peristaltic pump and the slurry was left at room temperature under ultrasonic mixing for 90 min. The precursor solution was prepared with iron (III) nitrate nonahydrate in order to obtain an iron load of 2 wt.%; the iron content in the catalysts should be very similar to the theoretical one, as verified in previous works using the same methodology [52]. After impregnation, the samples were dried at 110 °C overnight, heat treated at 400 °C during 1 h under nitrogen flow and reduced under hydrogen flow for 3 h at 400 °C.

The carbon-based materials were characterized by N_2 adsorption at −196 °C, by (EA, XPS, and by determination of pH_{PZC}. Further details about the techniques and conditions used can be found elsewhere [17].

3.3. Adsorption and Reaction Tests

The experiments were carried out in a stirred slurry batch reactor connected to a thermostatic bath (Hubber, polystat cc1) to maintain the temperature at 30 ± 0.5 °C. Immediately after reaching the desired temperature, 200 mL of PNP solution (3.6 mmol L^{-1}; a concentration typically found in industrial wastewater [29,53]) was adjusted to a pH of 3.0 (with 1 mol L^{-1} H_2SO_4); then 0.25 g L^{-1} of support was added, this being considered the time zero for the adsorption runs. For oxidation experiments, immediately after adding the catalyst (Fenton's oxidation) or support (CWPO), 29 mmol L^{-1} of 30% (w/v) H_2O_2 aqueous solution was also added, which corresponds to the initial reaction time (t = 0) of such runs. The initial pH and concentration of hydrogen peroxide were fixed in the best operating conditions obtained previously [17]. The oxidant concentration used is smaller than the stoichiometric amount required for total degradation of PNP (50 mmol L^{-1}) because the removal of PNP and formed intermediates/by-products is achieved by both adsorption and oxidation processes [17]. Constant stirring (at 200 rpm) was ensured by means of a bar and a magnetic plate (Falc).

Periodically, samples were taken from the reactor for measuring the PNP and total organic carbon (TOC) concentrations. Before analysis, the reaction was stopped by increasing the pH till 11 with 10 mol L^{-1} NaOH followed by neutralization (to pH ~ 7.0) with 1 mol L^{-1} H_2SO_4, then the carbon-based materials were removed by filtration with a 0.2 µm nylon syringe filter (from VWR). The residual hydrogen peroxide (after filtration) and iron (in case of Fenton's oxidation) concentrations in solution were measured, immediately (without stopping the reaction), at the end of the catalytic reactions runs. For assessing iron leaching, the samples were filtered through nylon filters with 0.45 μm of porosity and acidified until pH ~ 1.0 with concentrated nitric acid to keep the metal dissolved in solution.

The blank experiments—without PNP—were carried out with the goal of identifying the formation of hydroxyl radicals in solution; the tests proceeded in the same way as the catalytic runs, only the PNP solution was replaced by distilled water. Throughout the reactions, samples were taken and filtered, and then H_2O_2 concentration was measured and the presence of the hydroxyl radicals evaluated, as described in the following section.

3.4. *Analytical Methods*

Determination of the *p*-nitrophenol concentration and the evaluation of the presence of intermediate compounds, such as *p*-benzoquinone, hydroquinone, and *p*-nitrocatechol, were carried out by high pressure liquid chromatography (HPLC). TOC was measured in accordance with Method 5310 D [54] and was carried out in a Shimadzu TOC-L apparatus equipped with auto sampler (Shidmazu ASI-L). The hydrogen peroxide concentration was measured by the method developed by Sellers [55]. The formation of hydroxyl radicals was evaluated by the method proposed by Wang et al. [56]. The iron leached into the solution was measured by flame atomic absorption spectrometry (AAS)—Method 3111 B [54], using an AAS UNICAM spectrophotometer (model 939/959). The pH measurement was performed by potentiometric measurement (Method 4500 H^+ [54]). Further details about any of the analytical methods can be found elsewhere [17].

All analytical determinations were performed in duplicate and the coefficients of variation were less than 2% for all parameters.

4. Conclusions

Different undoped and N-doped carbon-based materials were prepared, characterized, and used as adsorbents or catalysts. In addition, these carbon-based materials were impregnated with iron (load of 2 wt.%) and used as catalysts in Fenton oxidation.

The textural properties of the carbon-based materials play an important role in the adsorption and/or catalytic activity. The support with the highest surface area -AC- presents the best performance during PNP adsorption, which is corroborated by the linear relationship between the PNP removal and the S_{BET} of the carbon-based materials. In contrast, the support or catalyst with the highest mesopores surface area -XG or Fe/XG- lead to the best removals by CWPO or Fenton's oxidation, respectively. On the other hand, the surface chemical properties also proved to be important in the performance of the materials, since the presence of N-functionalities, mainly pyrrole and pyridinic groups, on the carbons surface increases the PNP and TOC removals in all processes studied. A linear relationship between the PNP removal and N-content of carbon xerogels was obtained.

The Fe/XGM catalyst shows the best performances for the Fenton process with removals of 81% for PNP and 65% for TOC after 2 h. Reaction products identified (pyruvic, maleic, and oxalic acids) accounted, together with residual PNP, for more than 95% of the TOC along the experiments.

Stability and reusability of supports and catalysts were confirmed by recycling experiments. It was shown that the carbon-based materials were deactivated by hydrogen peroxide; in contrast, the iron catalysts were very stable and presented low coefficients of variation for PNP (1.4%) and TOC (0.3%) removals in the 3 cycles of reuse and without iron leaching.

The results obtained show that both textural and chemical properties of the carbon-based materials play a relevant role in the Fenton oxidation, where high mesoporosity combined with high surface area and N-functionalities improve their catalytic performance, demonstrating that carbon-based materials can be tailored to improve the efficiency of this advanced oxidation process.

Supplementary Materials: The following are available online at http://www.mdpi.com/2073-4344/9/3/258/s1, Figure S1: Correlation between specific surface area and the PNP removal after 2 h of adsorption with the carbon supports undoped (a) and N-doped with melamine (b) (pH = 3.0, T = 30 °C, [support] = 0.25 g L^{-1} and [PNP] = 3.6 mmol L^{-1}). Figure S2: PNP (a) and TOC (b) removal after 2 h of adsorption or Fenton's oxidation in presence of DMSO with the catalysts (pH = 3.0, T = 30 °C, [catalyst] = 0.25 g L^{-1}, $[H_2O_2]_{\text{when used}}$ = 29 mmol L^{-1}, [PNP] = 3.6 mmol L^{-1} and H_2O_2:PNP molar ratio = 8.1, when H_2O_2 was used). Figure S3: Relationship between PNP removal after 2 h of Fenton reaction and N-amount of N-doped xerogels (pH = 3.0, T = 30 °C, [support] = 0.25 g L^{-1}, $[H_2O_2]$ = 29 mmol L^{-1}, [PNP] = 3.6 mmol L^{-1} and H_2O_2:PNP molar ratio = 8.1).

Author Contributions: Conceptualization, methodology and investigation, O.S.G.P.S. and C.S.R.D.; resources, L.M.M. and M.F.R.P.; writing—original draft preparation, O.S.G.P.S. and C.S.R.D.; writing—review and editing, L.M.M. and M.F.R.P. All authors discussed the results and commented on the manuscript.

Funding: This research was partially funded by project "AIProcMat@N2020—Advanced Industrial Processes and Materials for a Sustainable Northern Region of Portugal 2020", with the reference NORTE-01-0145-FEDER-000006, supported by Norte Portugal Regional Operational Programme (NORTE 2020), under the Portugal 2020 Partnership Agreement, through the European Regional Development Fund (ERDF), by Projects UID/EQU/00511/2019—Laboratory for Process Engineering, Environment, Biotechnology and Energy—LEPABE and by Associate Laboratory LSRE-LCM—UID/EQU/50020/2019 funded by national funds through FCT/MCTES (PIDDAC).

Acknowledgments: Carmen Rodrigues is grateful to FCT for financial support through the Postdoctoral grant (SFRH/BPD/115879/2016) with financing from National and the European Social Funds through the Human Capital Operational Programme (POCH).

Conflicts of Interest: The authors declare no conflict of interest.

References

1. Von Sonntag, C. Advanced oxidation processes: Mechanistic aspects. *Water Sci. Technol.* **2008**, *58*, 1015–1021. [CrossRef] [PubMed]
2. Stasinakis, A.S. Use of selected advanced oxidation processes (aops) for wastewater treatment—A mini review. *Glob. NEST J.* **2008**, *10*, 376–385.
3. Laherty, K.A.; Huang, C.P. Continuous flow applications of fenton's reagent for the treatment of reafractory wastewater. In Proceedings of the Second International Symposium on Chemical Oxidation-Technologies for the Nineties, Nashville, TN, USA, 19–21 February 1992; CRC Press: Nashville, TN, USA, 1992.
4. Bigda, R.J. Consider fenton's chemistry for wastewater treatment. *Chem. Eng. Prog.* **1995**, *91*, 62–66.
5. Arslan, İ.; Akmehmet Balcioglu, I.; Tuhkanen, T. Oxidative treatment of simulated dyehouse effluent by uv and near-uv light assisted fenton's reagent. *Chemosphere* **1999**, *39*, 2767–2783. [CrossRef]
6. Papadopoulos, A.E.; Fatta, D.; Loizidou, M. Development and optimization of dark fenton oxidation for the treatment of textile wastewaters with high organic load. *J. Hazard. Mater.* **2007**, *146*, 558–563. [CrossRef] [PubMed]
7. Walling, C. Fenton's reagent revisited. *Acc. Chem. Res.* **1975**, *8*, 125–131. [CrossRef]
8. Queirós, S.; Morais, V.; Rodrigues, C.S.D.; Maldonado-Hódar, F.J.; Madeira, L.M. Heterogeneous fenton's oxidation using fe/zsm-5 as catalyst in a continuous stirred tank reactor. *Sep. Purif. Technol.* **2015**, *141*, 235–245. [CrossRef]
9. Navalon, S.; Alvaro, M.; Garcia, H. Heterogeneous fenton catalysts based on clays, silicas and zeolites. *Appl. Catal. B Environ.* **2010**, *99*, 1–26. [CrossRef]
10. Herney-Ramirez, J.; Vicente, M.A.; Madeira, L.M. Heterogeneous photo-fenton oxidation with pillared clay-based catalysts for wastewater treatment: A review. *Appl. Catal. B Environ.* **2010**, *98*, 10–26. [CrossRef]
11. Garrido-Ramírez, E.G.; Theng, B.K.G.; Mora, M.L. Clays and oxide minerals as catalysts and nanocatalysts in fenton-like reactions—A review. *Appl. Clay Sci.* **2010**, *47*, 182–192. [CrossRef]
12. Sashkina, K.A.; Polukhin, A.V.; Labko, V.S.; Ayupov, A.B.; Lysikov, A.I.; Parkhomchuk, E.V. Fe-silicalites as heterogeneous fenton-type catalysts for radiocobalt removal from edta chelates. *Appl. Catal. B Environ.* **2016**, *185*, 353–361. [CrossRef]
13. Esteves, B.M.; Rodrigues, C.S.D.; Boaventura, R.A.R.; Maldonado-Hódar, F.J.; Madeira, L.M. Coupling of acrylic dyeing wastewater treatment by heterogeneous fenton oxidation in a continuous stirred tank reactor with biological degradation in a sequential batch reactor. *J. Environ. Manag.* **2016**, *166*, 193–203. [CrossRef]

14. Duarte, F.; Maldonado-Hódar, F.J.; Madeira, L.M. New insight about orange ii elimination by characterization of spent activated carbon/fe fenton-like catalysts. *Appl. Catal. B Environ.* **2013**, *129*, 264–272. [CrossRef]
15. Menéndez, J.A.; Phillips, J.; Xia, B.; Radovic, L.R. On the modification and characterization of chemical surface properties of activated carbon: In the search of carbons with stable basic properties. *Langmuir* **1996**, *12*, 4404–4410. [CrossRef]
16. Auer, E.; Freund, A.; Pietsch, J.; Tacke, T. Carbons as supports for industrial precious metal catalysts. *Appl. Catal. A Gen.* **1998**, *173*, 259–271. [CrossRef]
17. Rodrigues, C.S.D.; Soares, O.S.G.P.; Pinho, M.T.; Pereira, M.F.R.; Madeira, L.M. P-nitrophenol degradation by heterogeneous fenton's oxidation over activated carbon-based catalysts. *Appl. Catal. B Environ.* **2017**, *219*, 109–122. [CrossRef]
18. Liu, X.; Yin, H.; Lin, A.; Guo, Z. Effective removal of phenol by using activated carbon supported iron prepared under microwave irradiation as a reusable heterogeneous fenton-like catalyst. *J. Environ. Chem. Eng.* **2017**, *5*, 870–876. [CrossRef]
19. Ramirez, J.H.; Maldonado-Hódar, F.J.; Pérez-Cadenas, A.F.; Moreno-Castilla, C.; Costa, C.A.; Madeira, L.M. Azo-dye orange ii degradation by heterogeneous fenton-like reaction using carbon-fe catalysts. *Appl. Catal. B Environ.* **2007**, *75*, 312–323. [CrossRef]
20. Messele, S.A.; Soares, O.S.G.P.; Órfão, J.J.M.; Bengoa, C.; Stüber, F.; Fortuny, A.; Fabregat, A.; Font, J. Effect of activated carbon surface chemistry on the activity of zvi/ac catalysts for fenton-like oxidation of phenol. *Catal. Today* **2015**, *240*, 73–79. [CrossRef]
21. Cleveland, V.; Bingham, J.-P.; Kan, E. Heterogeneous fenton degradation of bisphenol a by carbon nanotube-supported Fe_3O_4. *Sep. Purif. Technol.* **2014**, *133*, 388–395. [CrossRef]
22. Yao, Y.; Chen, H.; Lian, C.; Wei, F.; Zhang, D.; Wu, G.; Chen, B.; Wang, S. Fe, Co, Ni nanocrystals encapsulated in nitrogen-doped carbon nanotubes as fenton-like catalysts for organic pollutant removal. *J. Hazard. Mater.* **2016**, *314*, 129–139. [CrossRef]
23. Arshadi, M.; Abdolmaleki, M.K.; Mousavinia, F.; Khalafi-Nezhad, A.; Firouzabadi, H.; Gil, A. Degradation of methyl orange by heterogeneous fenton-like oxidation on a nano-organometallic compound in the presence of multi-walled carbon nanotubes. *Chem. Eng. Res. Des.* **2016**, *112*, 113–121. [CrossRef]
24. Carrasco-Díaz, M.R.; Castillejos-López, E.; Cerpa-Naranjo, A.; Rojas-Cervantes, M.L. On the textural and crystalline properties of fe-carbon xerogels. Application as fenton-like catalysts in the oxidation of paracetamol by H_2O_2. *Microporous Mesoporous Mater.* **2017**, *237*, 282–293. [CrossRef]
25. Pinho, M.T.; Gomes, H.T.; Ribeiro, R.S.; Faria, J.L.; Silva, A.M.T. Carbon nanotubes as catalysts for catalytic wet peroxide oxidation of highly concentrated phenol solutions: Towards process intensification. *Appl. Catal. B Environ.* **2015**, *165*, 706–714. [CrossRef]
26. Alegre, C.; Calvillo, L.; Moliner, R.; González-Expósito, J.A.; Guillén-Villafuerte, O.; Huerta, M.V.M.; Pastor, E.; Lázaro, M.J. Pt and ptru electrocatalysts supported on carbon xerogels for direct methanol fuel cells. *J. Power Sources* **2011**, *196*, 4226–4235. [CrossRef]
27. ATSDR. *Toxicological Profile for Nitrophenols: 2-Nitrophenol and 4-Nitrophenol*; Public Agency for Toxic Substances and Disease Registry, Health Service: Atlanta, GA, USA, 1992.
28. Tang, L.; Tang, J.; Zeng, G.; Yang, G.; Xie, X.; Zhou, Y.; Pang, Y.; Fang, Y.; Wang, J.; Xiong, W. Rapid reductive degradation of aqueous p-nitrophenol using nanoscale zero-valent iron particles immobilized on mesoporous silica with enhanced antioxidation effect. *Appl. Surf. Sci.* **2015**, *333*, 220–228. [CrossRef]
29. Ji, Q.; Li, J.; Xiong, Z.; Lai, B. Enhanced reactivity of microscale Fe/Cu bimetallic particles (Mfe/Cu) with persulfate (ps) for p-nitrophenol (pnp) removal in aqueous solution. *Chemosphere* **2017**, *172*, 10–20. [CrossRef]
30. Zhang, A.; Wang, N.; Zhou, J.; Jiang, P.; Liu, G. Heterogeneous fenton-like catalytic removal of p-nitrophenol in water using acid-activated fly ash. *J. Hazard. Mater.* **2012**, *201–202*, 68–73. [CrossRef]
31. Subbulekshmi, N.L.; Subramanian, E. Nano CuO immobilized fly ash zeolite fenton-like catalyst for oxidative degradation of p-nitrophenol and p-nitroaniline. *J. Environ. Chem. Eng.* **2017**, *5*, 1360–1371. [CrossRef]
32. Zhong, Y.; Liang, X.; He, Z.; Tan, W.; Zhu, J.; Yuan, P.; Zhu, R.; He, H. The constraints of transition metal substitutions (Ti, Cr, Mn, Co and Ni) in magnetite on its catalytic activity in heterogeneous fenton and uv/fenton reaction: From the perspective of hydroxyl radical generation. *Appl. Catal. B Environ.* **2014**, *150–151*, 612–618. [CrossRef]

33. Ferroudj, N.; Nzimoto, J.; Davidson, A.; Talbot, D.; Briot, E.; Dupuis, V.; Bée, A.; Medjram, M.S.; Abramson, S. Maghemite nanoparticles and maghemite/silica nanocomposite microspheres as magnetic fenton catalysts for the removal of water pollutants. *Appl. Catal. B Environ.* **2013**, *136–137*, 9–18. [CrossRef]
34. Wan, D.; Li, W.; Wang, G.; Lu, L.; Wei, X. Degradation of p-nitrophenol using magnetic fe0/fe3o4/coke composite as a heterogeneous fenton-like catalyst. *Sci. Total Environ.* **2017**, *574*, 1326–1334. [CrossRef]
35. Ribeiro, R.S.; Silva, A.M.T.; Figueiredo, J.L.; Faria, J.L.; Gomes, H.T. Removal of 2-nitrophenol by catalytic wet peroxide oxidation using carbon materials with different morphological and chemical properties. *Appl. Catal. B Environ.* **2013**, *140*, 356–362. [CrossRef]
36. Messele, S.A.; Soares, O.S.G.P.; Órfão, J.J.M.; Bengoa, C.; Font, J. Zero-valent iron supported on nitrogen-doped carbon xerogel as catalysts for the oxidation of phenol by fenton-like system. *Environ. Technol.* **2018**, *39*, 2951–2958. [CrossRef]
37. Dhaouadi, A.; Adhoum, N. Heterogeneous catalytic wet peroxide oxidation of paraquat in the presence of modified activated carbon. *Appl. Catal. B Environ.* **2010**, *97*, 227–235. [CrossRef]
38. Messele, S.A.; Soares, O.S.G.P.; Órfão, J.J.M.; Stüber, F.; Bengoa, C.; Fortuny, A.; Fabregat, A.; Font, J. Zero-valent iron supported on nitrogen-containing activated carbon for catalytic wet peroxide oxidation of phenol. *Appl. Catal. B Environ.* **2014**, *154*, 329–338. [CrossRef]
39. Yang, G.; Chen, H.; Qin, H.; Feng, Y. Amination of activated carbon for enhancing phenol adsorption: Effect of nitrogen-containing functional groups. *Appl. Surf. Sci.* **2014**, *293*, 299–305. [CrossRef]
40. Santos, V.P.; Pereira, M.F.R.; Faria, P.C.C.; Órfão, J.J.M. Decolourisation of dye solutions by oxidation with H_2O_2 in the presence of modified activated carbons. *J. Hazard. Mater.* **2009**, *162*, 736–742. [CrossRef]
41. Sun, H.; Kwan, C.; Suvorova, A.; Ang, H.M.; Tadé, M.O.; Wang, S. Catalytic oxidation of organic pollutants on pristine and surface nitrogen-modified carbon nanotubes with sulfate radicals. *Appl. Catal. B Environ.* **2014**, *154*, 134–141. [CrossRef]
42. Santos, D.F.M.; Soares, O.S.G.P.; Silva, A.M.T.; Figueiredo, J.L.; Pereira, M.F.R. Catalytic wet oxidation of organic compounds over n-doped carbon nanotubes in batch and continuous operation. *Appl. Catal. B Environ.* **2016**, *199*, 361–371. [CrossRef]
43. Liu, Y.; Wang, D.; Sun, B.; Zhu, X. Aqueous 4-nitrophenol decomposition and hydrogen peroxide formation induced by contact glow discharge electrolysis. *J. Hazard. Mater.* **2010**, *181*, 1010–1015. [CrossRef]
44. Sun, S.-P.; Lemley, A.T. P-nitrophenol degradation by a heterogeneous fenton-like reaction on nano-magnetite: Process optimization, kinetics, and degradation pathways. *J. Mol. Catal. A Chem.* **2011**, *349*, 71–79. [CrossRef]
45. Rodrigues, C.S.D.; Borges, R.A.C.; Lima, V.N.; Madeira, L.M. P-nitrophenol degradation by fenton's oxidation in a bubble column reactor. *J. Environ. Manag.* **2018**, *206*, 774–785. [CrossRef]
46. Wang, Y.; Wei, H.; Liu, P.; Yu, Y.; Zhao, Y.; Li, X.; Jiang, W.; Wang, J.; Yang, X.; Sun, C. Effect of structural defects on activated carbon catalysts in catalytic wet peroxide oxidation of m-cresol. *Catal. Today* **2015**, *258*, 120–131. [CrossRef]
47. Zazo, J.A.; Casas, J.A.; Mohedano, A.F.; Rodríguez, J.J. Catalytic wet peroxide oxidation of phenol with a fe/active carbon catalyst. *Appl. Catal. B Environ.* **2006**, *65*, 261–268. [CrossRef]
48. Ribeiro, R.S.; Fathy, N.A.; Attia, A.A.; Silva, A.M.T.; Faria, J.L.; Gomes, H.T. Activated carbon xerogels for the removal of the anionic azo dyes orange ii and chromotrope 2r by adsorption and catalytic wet peroxide oxidation. *Chem. Eng. J.* **2012**, *195–196*, 112–121. [CrossRef]
49. Gomes, H.T.; Miranda, S.M.; Sampaio, M.J.; Silva, A.M.T.; Faria, J.L. Activated carbons treated with sulphuric acid: Catalysts for catalytic wet peroxide oxidation. *Catal. Today* **2010**, *151*, 153–158. [CrossRef]
50. Sousa, J.P.S.; Pereira, M.F.R.; Figueiredo, J.L. Carbon xerogel catalyst for no oxidation. *Catalysts* **2012**, *2*, 447. [CrossRef]
51. Sousa, J.P.S.; Pereira, M.F.R.; Figueiredo, J.L. No oxidation over nitrogen doped carbon xerogels. *Appl. Catal. B Environ.* **2012**, *125*, 398–408. [CrossRef]
52. Soares, O.S.G.P.; Órfão, J.J.M.; Gallegos-Suarez, E.; Castillejos, E.; Rodriguez-Ramos, I.; Pereira, M.F.R. Nitrate reduction over a pd-cu/mwcnt catalyst: Application to a polluted groundwater. *Environ. Technol.* **2012**, *33*, 2353–2358. [CrossRef]
53. Bhatti, Z.I.; Toda, H.; Furukawa, K. P nitrophenol degradation by activated sludge attached on nonwovens. *Water Res.* **2002**, *36*, 1135–1142. [CrossRef]

54. APHA; AWWA; WEF. *Standard Methods for the Examination of Water and Wastewater*, 20th ed.; American Public Health Association, American Water Works Association, Water Pollution Control Federation: Washington, DC, USA, 1998.
55. Sellers, R.M. Spectrophotometric determination of hydrogen peroxide using potassium titanium (IV) oxalate. *Analyst* **1980**, *105*, 950–954. [CrossRef]
56. Wang, J.; Guo, Y.; Gao, J.; Jin, X.; Wang, Z.; Wang, B.; Li, K.; Li, Y. Detection and comparison of reactive oxygen species (Ros) generated by chlorophyllin metal (Fe, Mg and Cu) complexes under ultrasonic and visible-light irradiation. *Ultrason. Sonochem.* **2011**, *18*, 1028–1034. [CrossRef]

Article

Facile Synthesis of Bi_2MoO_6 Microspheres Decorated by CdS Nanoparticles with Efficient Photocatalytic Removal of Levfloxacin Antibiotic

Shijie Li [1,*], Yanping Liu [2], Yunqian Long [1,*], Liuye Mo [1], Huiqiu Zhang [1] and Jianshe Liu [3]

1 Key Laboratory of Key Technical Factors in Zhejiang Seafood Health Hazards, Institute of Innovation & Application, Zhejiang Ocean University, Zhoushan 316022, China; liuyemo@zjou.edu.cn (L.M.); zhanghuiqiu2006@163.com (H.Z.)

2 Department of Environmental Engineering, Zhejiang Ocean University, Zhoushan 316022, China; liuyp@zjou.edu.cn

3 State Environmental Protection Engineering Center for Pollution Treatment and Control in Textile Industry, College of Environmental Science and Engineering, Donghua University, Shanghai 201620, China; liujianshe@dhu.edu.cn

* Correspondence: lishijie@zjou.edu.cn (S.L.); longyunqian@163.com (Y.L.); Tel.: +86-21-67792557 (S.L.)

Received: 26 September 2018; Accepted: 18 October 2018; Published: 19 October 2018

Abstract: Developing high-efficiency and stable visible-light-driven (VLD) photocatalysts for removal of toxic antibiotics is still a huge challenge at present. Herein, a novel CdS/Bi_2MoO_6 heterojunction with CdS nanoparticles decorated Bi_2MoO_6 microspheres has been obtained by a simple solvothermal-precipitation-calcination method. 1.0CdS/Bi_2MoO_6 has stronger light absorption ability and highest photocatalytic activity with levofloxacin (LEV) degradation efficiency improving 6.2 or 12.6 times compared to pristine CdS or Bi_2MoO_6. CdS/Bi_2MoO_6 is very stable during cycling tests, and no appreciable activity decline and microstructural changes are observed. Results signify that the introduction of CdS could enhance the light absorption ability and dramatically boost the separation of charge carriers, leading to the excellent photocatalytic performance of the heterojunction. This work demonstrates that flower-like CdS/ Bi_2MoO_6 is an excellent photocatalyst for the efficient removal of the LEV antibiotic.

Keywords: CdS; Bi_2MoO_6 microspheres; antibiotic removal; charge separation; visible-light-driven

1. Introduction

Antibiotics accumulated in water have posed a great threat to the environmental safety and human health. In particular, levofloxacin (LEV), as a typical fluoroquinolone antibiotic, has been widely used to treat human and animal infections. As a result, LEV was excessively accumulated in aquatic environments, causing serious health-risk issues by inducing genetic exchange and the bacterial drug resistance [1,2]. Conventional technologies for the removal of antibiotics in aquatic systems have been restricted due to low efficiency, high cost and/or sophisticated instrumentation. In recent years, the semiconductor-based photocatalysis as a green oxidation technology, has made significant developments for treating water pollutions [3–10]. Bismuth (III)-based semiconductors (e.g., BiOX (X = Cl, Br, I) [11–13], Bi_2MoO_6 [14–17], and Bi_2WO_6 [18]) have been emerging as kinds of high-efficiency photocatalysts. Among them, Bi_2MoO_6 as a good visible-light-driven (VLD) photocatalyst, has been one of the focal research points in the area of photocatalysis. Unlike the popular TiO_2, which only can respond to the ultraviolet light, Bi_2MoO_6 (Eg $\approx$ 2.5–2.8 eV) is active in the visible-light region. However, Bi_2MoO_6 still suffers from the narrow range of sunlight response and fast electron-hole recombination [15–17,19]. Intriguingly, the fabrication of semiconductor heterojunctions is a crucial way to obtain superior photocatalysts [20]. Thus, a multiple of Bi_2MoO_6-based heterojunctions have been fabricated to ameliorate the photocatalytic activity for the removal of pollutants, such as BN/Bi_2MoO_6 [21], Bi/Bi_2MoO_6 [22], C_3N_4/Bi_2MoO_6 [23,24], TiO_2/Bi_2MoO_6 [25], etc. [26–31]. Our group has also developed some Bi_2MoO_6-based heterojunction photocatalysts with excellent photocatalytic properties [32–36]. However, the majority of them were employed to degrade traditional dyes to evaluate their activity, their application in decomposing colorless and refractory pollutants were underdeveloped.

CdS is known as a promising VLD photocatalyst in virtue of its favorable visible-light-response, and chemical stability [37–40]. Due to the well-matched band structures of CdS and Bi_2MoO_6 [31], it is anticipated the rationally designed CdS/Bi_2MoO_6 can be endowed with superior photocatalytic activity for the removal of LEV antibiotic. However, reports on photocatalytic degradation of LEV antibiotic using flower-like CdS/Bi_2MoO_6 have not been indicated.

Inspired by the above aspects, we designed a flower-like CdS/Bi_2MoO_6 for the removal of the LEV antibiotic. A facile solvothermal-precipitation-calcination method was employed to construct hierarchical CdS/Bi_2MoO_6 heterojunctions with tightly interfacial contact. 1.0CdS/Bi_2MoO_6 possesses the highest activity towards the removal of LEV antibiotic. The plausible photocatalytic mechanism for the degradation of LEV antibiotic over CdS/Bi_2MoO_6 was also illustrated.

2. Results

2.1. Preparation and Characterization

CdS/Bi_2MoO_6 heterojunctions with CdS/Bi_2MoO_6 molar ratios of 0.5:1, 0.75:1, 1.0:1, and 1.5:1 were fabricated and denoted as 0.5CdS/Bi_2MoO_6, 0.75CdS/Bi_2MoO_6, 1.0CdS/Bi_2MoO_6, and 1.5CdS/Bi_2MoO_6, respectively. The crystalline phases of Bi_2MoO_6, CdS, and the optimal sample (1.0CdS/Bi_2MoO_6) were examined by XRD technique (Figure 1). All the diffraction peaks for Bi_2MoO_6 and CdS correspond to crystal planes of orthorhombic (JCPDS 76-2388) [25,33] and (JCPDS 75-1545) [31], respectively. As for 1.0 CdS/Bi_2MoO_6, both the diffraction peaks belonging to Bi_2MoO_6 and CdS were detected in the XRD pattern, verifying the fabrication of CdS/Bi_2MoO_6.

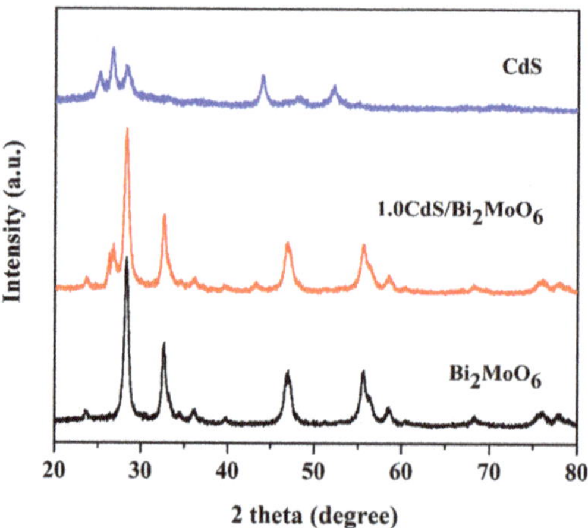

Figure 1. XRD patterns of Bi_2MoO_6, CdS, and 1.0CdS/Bi_2MoO_6.

The morphology of Bi_2MoO_6 and 1.0CdS/Bi_2MoO_6 was characterized by SEM and TEM analysis. As shown in Figure S1, pristine Bi_2MoO_6 has the shape of a microsphere with a smooth surface, in accordance with the morphology of the reported Bi_2MoO_6 [25]. In comparison with Bi_2MoO_6, the surface of 1.0CdS/Bi_2MoO_6 became rough due to the coating of CdS nanoparticles (size: 5–25 nm) (Figure 2a,b), such a hetero-structure favors the separation of charges. In addition, the EDX spectrum shows the co-existence of Cd, S, Bi, Mo and O elements in 1.0CdS/Bi_2MoO_6 (Figure 2c). The detailed microstructure of 1.0CdS/Bi_2MoO_6 revealed by TEM analysis (Figure 2d) shows that the surfaces of Bi_2MoO_6 nanoplates (size: 200 nm) are deposited by numerous of CdS nanoparticles (size: 5–25 nm). The above characterizations further verify the formation of CdS/Bi_2MoO_6.

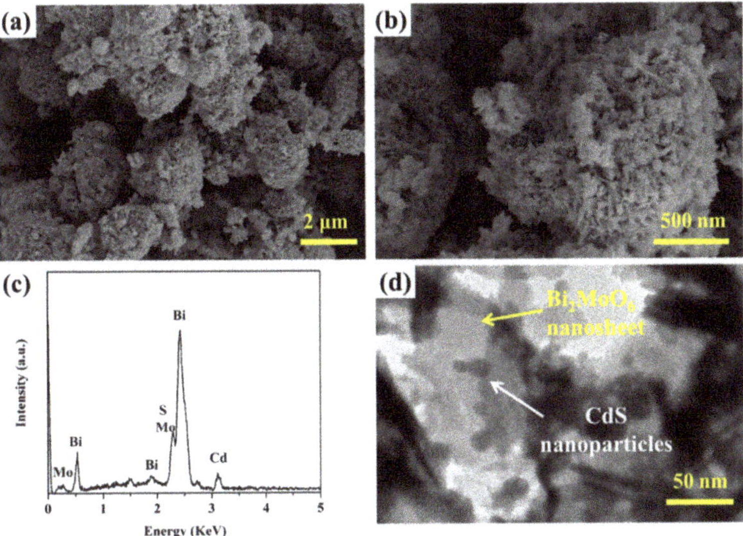

Figure 2. (**a**,**b**) SEM images, (**c**) EDX spectrum, and (**d**) TEM image of 1.0CdS/Bi_2MoO_6.

The UV-Vis absorption spectra of Bi_2MoO_6, CdS, and 1.0CdS/Bi_2MoO_6 are illustrated in Figure 3. Pure Bi_2MoO_6 and CdS exhibit good absorption in the visible light region, consistent with the previous reports [18,37,40]. Notably, 1.0CdS/Bi_2MoO_6 presents better optical absorption than Bi_2MoO_6, indicating that 1.0CdS/Bi_2MoO_6 possesses a better ability to harvest sunlight.

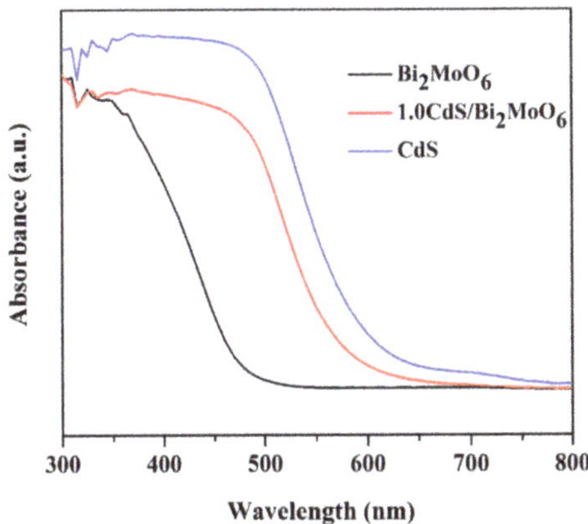

Figure 3. UV-vis absorbance spectra of bare Bi_2MoO_6, CdS, and 1.0CdS/Bi_2MoO_6.

2.2. Photocatalytic Performance

A typical fluoroquinolone antibiotic LEV is a kind of persistent organic pollutants (POPS) due to its water solubility and chemical stability. Photocatalysis is a promising technique for the removal of LEV antibiotic. For example, Kaur et al. fabricated Bi_2WO_6 nanocuboid (0.75 g/L) for almost

80% degradation of LEV (10 mg/L) within 150 min under visible light irradiation [41]. In this work, the photocatalytic activity of CdS/Bi_2MoO_6 heterojunctions was also evaluated by degrading LEV as a representative toxic antibiotic under visible light (Figure 4). No LEV degradation was detected in the blank test (without catalysts). Pure CdS or Bi_2MoO_6 shows very limited photocatalytic activity, as only 20.6% or 10.8% of LEV was decomposed within 60 min of reaction, owing to the fast recombination of electron-hole pairs. Encouragingly, with the introduction of CdS into the composites, all of the CdS/Bi_2MoO_6 heterojunctions show much enhanced photocatalytic activity compared to pristine CdS and Bi_2MoO_6. Notably, 1.0 CdS/Bi_2MoO_6 obtains the highest activity with 80.4% of LEV degradation within 60 min, much higher than CdS (20.6%), Bi_2MoO_6 (10.9%), or the mechanical mixture (25.8%), verifying the existence of synergistic interaction between CdS and Bi_2MoO_6. Since the introduction of CdS nanoparticles on Bi_2MoO_6 microspheres, the intimately contacted interfacial surface is formed for effective separation of charge carriers. However, with the excessive deposition of CdS ($1.5CdS/Bi_2MoO_6$), the agglomeration of CdS undermines the synergistic effect, and thus suppresses the interfacial charge transfer, resulting in the decrease of the photocatalytic activity.

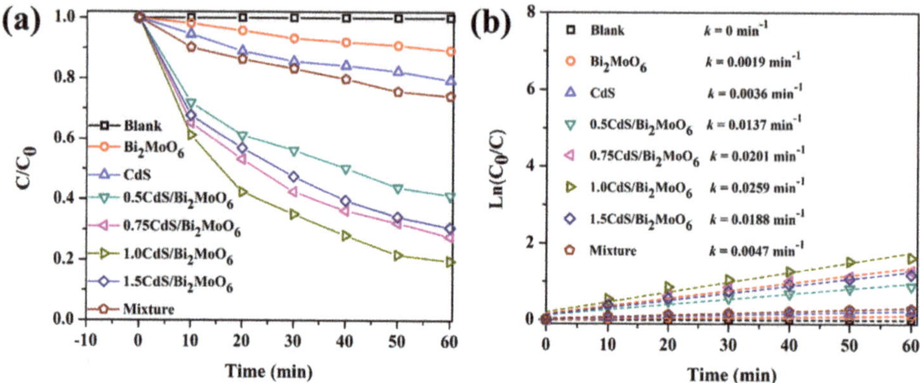

Figure 4. (**a**) The photo-degradation efficiency against LEV (20 mg L^{-1}, 100 mL) by different catalysts (35 mg). (**b**) Kinetic modeling for LEV removal over different catalysts.

To make a quantitative comparison, the LEV degradation data were fitted with the pseudo-first-order model (Figure 4b), $-\ln(C/C_0) = kt$. Clearly, $1.0CdS/Bi_2MoO_6$ has the largest rate constant of 0.0259 min^{-1}, which is approximately 6.2, 12.6 or 4.5 times higher than pristine CdS (0.0036 min^{-1}), Bi_2MoO_6 (0.0019 min^{-1}), and the mixture (0.0047 min^{-1}).

The mineralization of organic contaminants is pivotal in the wastewater treatment. To assess the mineralization of LEV, TOC values were recorded during the degradation of LEV (40 mg L^{-1}, 200 mL) by $1.0CdS/Bi_2MoO_6$ (200 mg). As depicted in Figure 5, 90.1% of TOC was removed after 2.5 h of reaction. These results indicate that $1.0CdS/Bi_2MoO_6$ could effectively decompose and mineralize the LEV antibiotic.

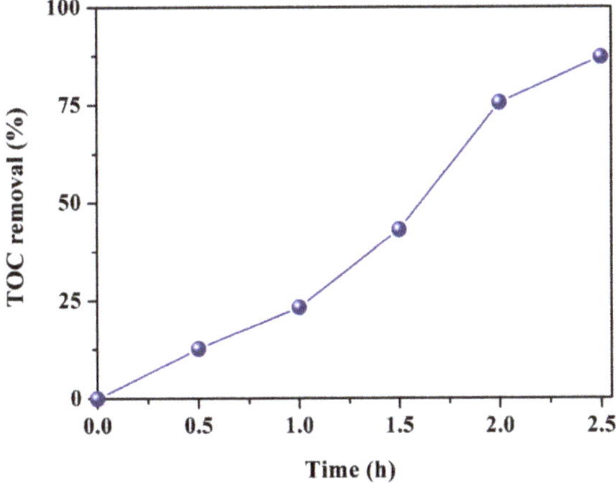

Figure 5. TOC removal profile of LEV (40 mg L^{-1}, 200 mL) over 1.0CdS/Bi_2MoO_6 (200 mg).

The reusability of a photocatalyst is an important factor for the practical treatment of wastewater. Therefore, the cycling runs in LEV degradation over 1.0CdS/Bi_2MoO_6 were executed. Inspiringly, 1.0CdS/Bi_2MoO_6 retained its initial photocatalytic activity after six successive runs (Figure 6a). Moreover, compared to the crystal phase of the fresh 1.0CdS/Bi_2MoO_6, no obvious phase changes (Figure 6b) of the used one was detected, illustrating that 1.0CdS/Bi_2MoO_6 has superior stability and reusability.

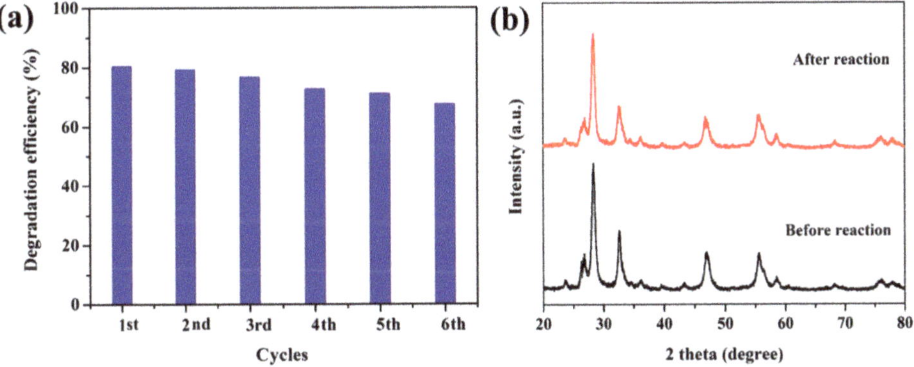

Figure 6. (**a**) Recycling tests of 1.0CdS/Bi_2MoO_6 for LEV degradation; (**b**) XRD patterns of 1.0CdS/Bi_2MoO_6 before and after six cycles.

2.3. Photocatalytic Reaction Mechanism

A batch of experiments was conducted to analyze the photocatalytic reaction mechanism accounting for the degradation of the LEV antibiotic under visible light through adding various scavengers (Figure 7). The LEV degradation rate showed no obvious decline when IPA (scavenger of •OH) was introduced, indicating that •OH did not play a pivotal role. By contrast, the LEV degradation rate was substantially suppressed by BQ (quencher of $O_2^{\bullet-}$) or AO (quencher of h^+), reflecting that $O_2^{\bullet-}$ and h^+ could be the dominant active species responsible for the LEV degradation over 1.0CdS/Bi_2MoO_6.

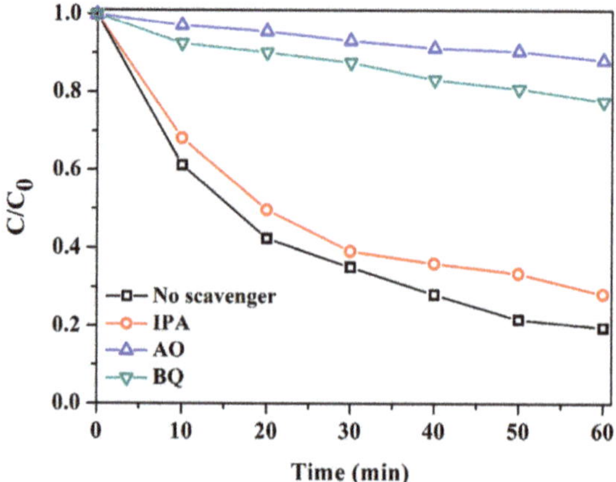

Figure 7. Photo-degradation of LEV (20 mg L^{-1}, 100 mL) by1.0CdS/Bi_2MoO_6 (35 mg) under visible light in the presence of isopropyl alcohol (IPA), ammonium oxalate (AO), and benzoquinone (BQ).

The bandgap (E_g) of Bi_2MoO_6 and CdS were calculated by the equation: $(\alpha h\nu) = A(h\nu\text{-}E_g)^{n/2}$. As presented in Figure S2, the E_g of Bi_2MoO_6 and CdS are about 2.66 eV and 2.24 eV, consistent with the reference [32,40,42]. The valence band (VB) and conduction band (CB) of Bi_2MoO_6 and CdS can be obtained by the following equation: $E_{VB} = X - E_0 + 0.5\ E_g$ (1), $E_{CB} = E_{VB} - E_g$ (2). Thereby, the E_{VB} and E_{CB} of Bi_2MoO_6 were −0.32 and 2.34 eV [36], while those of CdS were −0.56 and 1.68 eV [40]. Apparently, a staggered type II band structure can be constructed in CdS /Bi_2MoO_6, beneficial to retarding the recombination of charge carriers.

Since the separation and transport rate of electron-hole pairs of a semiconductor are closely related to the fluorescence emission [29,43–45], the photoluminescence (PL) technique was applied to study the transport behaviors of charge carriers in Bi_2MoO_6 and 1.0CdS/Bi_2MoO_6 (Figure S3). Through comparing the PL emission spectra of Bi_2MoO_6 and 1.0CdS/Bi_2MoO_6 (Figure S3), it is found that 1.0CdS/Bi_2MoO_6 exhibits much lower fluorescence intensity compared to that of pure Bi_2MoO_6, indicating that the nano-junction between Bi_2MoO_6 and CdS greatly promotes the separation of electron-hole pairs.

On the basis of above experimental results, the interfacial charge transfer behavior of CdS/Bi_2MoO_6 is illustrated in Figure 8. Under visible-light illumination, electron-hole pairs can be produced in both CdS and Bi_2MoO_6. The photo-excited e^- in the CB of CdS may rapidly migrate to that of Bi_2MoO_6, while the photo-excited holes in the VB of Bi_2MoO_6 can preferably drift to that of CdS. Such an interfacial charge movement can effectively retard the electron-hole recombination, accounting for the amelioration of photocatalytic performance of CdS/Bi_2MoO_6. The CB potential of CdS (−0.56 eV) and is Bi_2MoO_6 (−0.32 eV) are more negative than E ($O_2/O_2^{\bullet-}$) (+0.13 eV) [46], thus the e^- in the CB of CdS and Bi_2MoO_6 can react with O_2 to generate active $O_2^{\bullet-}$, further decomposing LEV antibiotic. On the other hand, the holes collected in the VB of CdS and Bi_2MoO_6 can directly decompose the LEV antibiotic due to their strong oxidation ability.

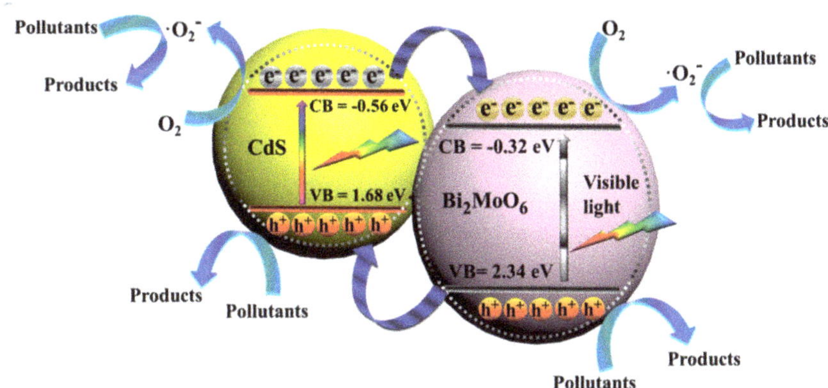

Figure 8. Photocatalytic mechanism scheme over CdS/Bi_2MoO_6.

3. Materials and Methods

3.1. Chemicals

All reagents were purchased from Shanghai Chemical Reagent factory (China) and used directly.

3.2. Chemicals Synthesis of Catalysts

Synthesis of Bi_2MoO_6: Typically, 1 mmol $Bi(NO_3)_3·5H_2O$ and 0.5 mmol $Na_2MoO_4·2H_2O$ were added into 50 mL of ethylene glycol, and the mixture turned into clear solution with the assistance of ultrasonication for 0.5 h. After that, 30 mL of ethanol was poured into the above solution and kept stirring for 0.5 h. Subsequently, the resulting solution transferred into an autoclave with the volume of 100 mL and reacted at 160 °C for 20 h in an oven.

Synthesis of CdS/Bi_2MoO_6: 0.5 mmol Bi_2MoO_6 was suspended into 50 mL of deionized water. Then, 1mmol $CdCl_2·2.5H_2O$ was dissolved in the above suspension under vigorously stirring. After that, 40 mL of Na_2S (1 mmol) aqueous solution was added into the above system dropwise under magnetically stirring for 2 h. The precipitation labeled as 1.0 CdS/Bi_2MoO_6 was washed, dried and then calcined at 200 °C for 2 h in N_2 atmosphere to obtain the catalysts. Through adjusting the amount of precursor of CdS, the samples with CdS/Bi_2MoO_6 molar ratios of 0.5:1, 0.75:1, and 1.5:1 are labeled as $0.5CdS/Bi_2MoO_6$, $0.75CdS/Bi_2MoO_6$, and $1.5CdS/Bi_2MoO_6$, respectively. The CdS sample was also prepared by the same procedure in the absence of Bi_2MoO_6.

3.3. Characterization

The phases of products were characterized via X-ray diffractometer (XRD, Bruker D8 ADVANCE, Karlsruhe, Germany) equipped with mono-chromatized Cu Kα radiation. UV-Vis diffuse reflectance spectra (DRS) of products were collected using Shimadzu UV-2600 spectrophotometer with BaSO4 as the reference standard. Scanning electron microscopy (SEM, Hitachi S-4800, Tokyo, Japan) and transmission electron microscopy (TEM, JEM-2100 JEOL, Tokyo, Japan) were employed to characterize the microstructures of the samples. The corresponding elemental components were analyzed by energy-dispersive X-ray spectroscopy (EDS, Bruker Quantax 400, Berlin, Germany). Photoluminescence (PL) spectra were conducted using Hitachi RF-6000 spectrophotofluorometer with the excitation wavelength of 300 nm.

3.4. Photocatalytic Performance Tests

The photocatalytic property of CdS/Bi_2MoO_6 was assessed by degradation of LEV antibiotic under visible-light illumination, and the light source is provided by a 300 W xenon lamp with a light filter $\lambda > 400$ nm. 50 mg of catalyst was scattered in LEV (100 mL, 20 mg L^{-1}) solution. The suspension was stirred in the dark for 0.5 h. Then the reaction was initiated when the lamp switches on. During the irradiation, 1.5 mL of suspension was sampled in 10 min. The LEV concentrations were analyzed by using UV-2600 spectrophotometer. Total organic carbon (TOC) of the LEV solutions during reaction was determined by using a Shimadzu TOC analyzer.

4. Conclusions

In summary, CdS nanoparticles interspersed-Bi_2MoO_6 heterojunction photocatalytst was fabricated by a facile solvothermal-precipitation-calcination method. Due to the introduction of CdS, the photo-absorption of CdS/Bi_2MoO_6 and the interfacial charge separation were remarkably enhanced and promoted. By virtue of these benefits, in comparison with bare CdS and Bi_2MoO_6, 1.0CdS/Bi_2MoO_6 presents profoundly enhanced visible-light photocatalytic activity for the removal of the LEV antibiotic. Moreover, 1.0CdS/Bi_2MoO_6 also possesses good stability and can effectively mineralize the LEV antibiotic. This work offers new insight into the design of high-performance Bi-based heterojunction photocatalysts for antibiotic wastewater treatment.

Supplementary Materials: The following are available online at http://www.mdpi.com/2073-4344/8/10/477/s1, Figure S1: SEM image of pure Bi_2MoO_6, Figure S2: The resulting Tauc plots of CdS and Bi_2MoO_6, Figure S3: Photoluminescence (PL) spectra of bare Bi_2MoO_6 and 1.0CdS/Bi_2MoO_6.

Author Contributions: S.L.: idea, and design of the paper; performed the tests; analyzed the data; wrote this paper. Y.L., Y.L., H.Z., L.M., and J.L. assisted the characterizations.

Funding: This work has been financially supported by the National Natural Science Foundation of China (51708504), the Public Projects of Zhejiang Province (LGN18E080003), the Science and Technology Project of Zhoushan (2017C41006), and the Qingdao Science and Technology Program (16-5-1-32-jch).

Conflicts of Interest: The authors declare no conflict of interest.

References

1. Du, J.; Guo, W.; Wang, H.; Yin, R.; Zheng, H.; Feng, X.; Che, D.; Ren, N. Hydroxyl radical dominated degradation of aquatic sulfamethoxazole by Fe^0/bisulfite/O_2: Kinetics, mechanisms, and pathways. *Water Res.* **2018**, *138*, 323–332. [CrossRef]
2. Hong, Y.Z.; Li, C.S.; Zhang, G.Y.; Meng, Y.D.; Yin, B.X.; Zhao, Y.; Shi, W.D. Efficient and stable Nb_2O_5 modified g-C_3N_4 photocatalyst for removal of antibiotic pollutant. *Chem. Eng. J.* **2016**, *299*, 74–84. [CrossRef]
3. Wang, F.; Li, Q.; Xu, D. Recent progress in semiconductor-based nanocomposite photocatalysts for solar-to-chemical energy conversion. *Adv. Energy Mater.* **2017**, *7*, 1700529. [CrossRef]
4. Cates, E.L. Photocatalytic water treatment: So where are we going with this? *Environ. Sci. Technol.* **2017**, *51*, 757–758. [CrossRef] [PubMed]
5. Pirhashemi, M.; Habibi-Yangjeh, A.; Pouran, S.R. Review on the criteria anticipated for the fabrication of highly efficient ZnO-based visible-light-driven photocatalysts. *J. Ind. Eng. Chem.* **2018**, *62*, 1–25. [CrossRef]
6. Zhu, S.S.; Wang, D.W. Photocatalysis: Basic principles, diverse forms of implementations and emerging scientifc opportunities. *Adv. Energy Mater.* **2017**, *7*, 1700841. [CrossRef]
7. Li, S.; Hu, S.; Jiang, W.; Liu, Y.; Liu, Y.; Zhou, Y.; Mo, L.; Liu, J. Ag_3VO_4 nanoparticles decorated $Bi_2O_2CO_3$ micro-flowers: An efficient visible-light-driven photocatalyst for the removal of toxic contaminants. *Front. Chem.* **2018**, *6*, 255. [CrossRef] [PubMed]
8. Li, S.; Hu, S.; Jiang, W.; Liu, Y.; Liu, J.; Wang, Z. Facile synthesis of flower-like Ag_3VO_4/Bi_2WO_6 heterojunction with enhanced visible-light photocatalytic activity. *J. Colloid Interface Sci.* **2017**, *501*, 156–163. [CrossRef] [PubMed]
9. Wang, Y.; Wang, H.; Xu, A.; Song, Z. Facile synthesis of Ag_3PO_4 modified with GQDs composites with enhanced visible-light photocatalytic activity. *J. Mater. Sci. Mater. Electron.* **2018**, *29*, 16691–16701. [CrossRef]
10. Zheng, J.; Chang, F.; Jiao, M.; Xu, Q.; Deng, B.; Hu, X. A visible-light-driven heterojuncted composite WO_3/$Bi_{12}O_{17}Cl_2$: Synthesis, characterization, and improved photocatalytic performance. *J. Colloid Interface Sci.* **2018**, *510*, 20–31. [CrossRef] [PubMed]
11. Cheng, H.F.; Huang, B.B.; Dai, Y. Engineering BiOX (X = Cl, Br, I) nanostructures for highly efficient photocatalytic applications. *Nanoscale* **2014**, *6*, 2009–2026. [CrossRef] [PubMed]

12. Yang, Y.; Zhang, C.; Lai, C.; Zeng, G.; Huang, D.; Cheng, M.; Wang, J.; Chen, F.; Zhou, C.; Xiong, W. BiOX (X = Cl, Br, I) photocatalytic nanomaterials: Applications for fuels and environmental management. *Adv. Colloid Interface Sci.* **2018**, *254*, 76–93. [CrossRef] [PubMed]
13. Li, S.J.; Hu, S.W.; Xu, K.B.; Jiang, W.; Liu, J.S.; Wang, Z.H. A novel heterostructure of BiOI nanosheets anchored onto MWCNTs with excellent visible-light photocatalytic activity. *Nanomaterials* **2017**, *7*, 22–34. [CrossRef] [PubMed]
14. Yu, C.; Wu, Z.; Liu, R.; Dionysiou, D.D.; Yang, K.; Wang, C.; Liu, H. Novel fluorinated Bi_2MoO_6 nanocrystals for efficient photocatalytic removal of water organic pollutants under different light source illumination. *Appl. Catal. B* **2017**, *209*, 1–11. [CrossRef]
15. Jia, Y.; Ma, Y.; Tang, J.; Shi, W. Hierarchical nanosheet-based Bi_2MoO_6 microboxes for efficient photocatalytic performance. *Dalton Trans.* **2018**, *47*, 5542–5547. [CrossRef] [PubMed]
16. Chen, Y.; Yang, W.; Gao, S.; Zhu, L.; Sun, C.; Li, Q. Internal polarization modulation in Bi_2MoO_6 for photocatalytic performance enhancement under visible light illumination. *ChemSusChem* **2018**, *11*, 1521–1532. [CrossRef] [PubMed]
17. Long, J.L.; Wang, S.C.; Chang, H.J.; Zhao, B.Z.; Liu, B.T.; Zhou, Y.G.; Wei, W.; Wang, X.X.; Huang, L.; Huang, W. Bi_2MoO_6 nanobelts for crystal facet-enhanced photocatalysis. *Small* **2014**, *10*, 2791–2795. [CrossRef] [PubMed]
18. Wang, J.; Tang, L.; Zeng, G.; Liu, Y.; Zhou, Y.; Deng, Y.; Wang, J.; Peng, B. Plasmonic Bi metal deposition and g-C_3N_4 coating on Bi_2WO_6 microspheres for efficient visible-light photocatalysis. *ACS Sustain. Chem. Eng.* **2017**, *5*, 1062–1072. [CrossRef]
19. Guo, C.; Xu, J.; Wang, S.; Li, L.; Zhang, Y.; Li, X. Facile synthesis and photocatalytic application of hierarchical mesoporous Bi_2MoO_6 nanosheet-based microspheres. *CrystEngComm* **2012**, *14*, 3602–3608. [CrossRef]
20. Ye, R.; Zhao, J.; Wickemeyer, B.B.; Toste, F.D.; Somorjai, G.A. Foundations and strategies of the construction of hybrid catalysts for optimized performances. *Nat. Catal.* **2018**, *1*, 318–325. [CrossRef]
21. Ke, J.; Duan, X.; Luo, S.; Zhang, H.; Sun, H.; Liu, J.; Tade, M.; Wang, S. UV-assisted construction of 3D hierarchical rGO/Bi_2MoO_6 composites for enhanced photocatalytic water oxidation. *Chem. Eng. J.* **2017**, *313*, 1447–1453. [CrossRef]
22. Zhao, Z.W.; Zhang, W.; Sun, Y.J.; Yu, J.Y.; Zhang, Y.X.; Wang, H.; Dong, F.; Wu, Z.B. Bi cocatalyst/Bi_2MoO_6 microspheres nanohybrid with SPR-promoted visible-light photocatalysis. *J. Phys. Chem. C* **2016**, *120*, 11889–11898. [CrossRef]
23. Opoku, F.; Govender, K.; Sittert, C.G.C.E.V.; Govender, P. Insights into the photocatalytic mechanism of mediator-free direct Z-scheme g-C_3N_4/Bi_2MoO_6 (010) and g-C_3N_4/Bi_2WO_6 (010) heterostructures: A hybrid density functional theory study. *Appl. Surf. Sci.* **2018**, *427*, 487–498. [CrossRef]
24. Ma, T.J.; Wu, J.; Mi, Y.D.; Chen, Q.H.; Ma, D.; Chai, C. Novel Z-Scheme g-C_3N_4/C@Bi_2MoO_6 composite with enhanced visible-light photocatalytic activity for naphthol degradation. *Sep. Purif. Technol.* **2017**, *183*, 54–65. [CrossRef]
25. Zhang, M.Y.; Shao, C.L.; Mu, J.B.; Zhang, Z.Y.; Guo, Z.C.; Zhang, P.; Liu, Y.C. One-dimensional Bi_2MoO_6/TiO_2 hierarchical heterostructures with enhanced photocatalytic activity. *CrystEngComm* **2012**, *14*, 605–612. [CrossRef]
26. Lv, J.; Zhang, J.; Liu, J.; Li, Z.; Dai, K.; Liang, C. Bi SPR-promoted Z-scheme Bi_2MoO_6/CdS-diethylenetriamine composite with effectively enhanced visible light photocatalytic hydrogen evolution activity and stability. *ACS Sustain. Chem. Eng.* **2018**, *6*, 696–706. [CrossRef]
27. Meng, Q.Q.; Zhou, Y.S.; Chen, G.; Hu, Y.D.; Lv, C.; Qiang, L.S.; Xing, W.N. Integrating both homojunction and heterojunction in QDs self-decorated Bi_2MoO_6/BCN composites to achieve an efficient photocatalyst for Cr(VI) reduction. *Chem. Eng. J.* **2018**, *334*, 334–343. [CrossRef]
28. Li, X.; Su, M.; Zhu, G.; Zhang, K.; Zhang, X.; Fan, J. Fabrication of a novel few-layer WS_2/Bi_2MoO_6 plate-on-plate heterojunction structure with enhanced visible-light photocatalytic activity. *Dalton Trans.* **2018**, *47*, 10046–10056. [CrossRef] [PubMed]
29. Zhang, J.L.; Ma, Z. Flower-like Ag_2MoO_4/Bi_2MoO_6 heterojunctions with enhanced photocatalytic activity under visible light irradiation. *J. Taiwan Inst. Chem. Eng.* **2017**, *71*, 156–164. [CrossRef]
30. Zhang, J.L.; Zhang, L.S.; Yu, N.; Xu, K.B.; Li, S.J.; Wang, H.L.; Liu, J.S. Flower-like Bi_2S_3/Bi_2MoO_6 heterojunction superstructures with enhanced visible-light-driven photocatalytic activity. *RSC Adv.* **2015**, *5*, 75081–75088. [CrossRef]

31. Feng, Y.; Yan, X.; Liu, C.; Hong, Y.; Zhu, L.; Zhou, M.; Shi, W. Hydrothermal synthesis of CdS/Bi2MoO6 heterojunction photocatalysts with excellent visible-light-driven photocatalytic performance. *Appl. Surf. Sci.* **2015**, *353*, 87–94. [CrossRef]
32. Li, S.; Jiang, W.; Hu, S.; Liu, Y.; Liu, Y.; Xu, K.; Liu, J. Hierarchical heterostructure of Bi_2MoO_6 micro-flowers decorated with Ag_2CO_3 nanoparticles for efficient visible-light-driven photocatalytic removal of toxic pollutants. *Beilstein J. Nanotechnol.* **2018**, *9*, 2297–2305. [CrossRef] [PubMed]
33. Li, S.; Hu, S.; Jiang, W.; Zhou, Y.; Liu, J.; Wang, Z. Facile synthesis of cerium oxide nanoparticles decorated flower-like bismuth molybdate for enhanced photocatalytic activity toward organic pollutant degradation. *J. Colloid Interface Sci.* **2018**, *530*, 171–178. [CrossRef] [PubMed]
34. Li, S.; Hu, S.; Jiang, W.; Liu, Y.; Zhou, Y.; Liu, Y.; Mo, L. Hierarchical architectures of bismuth molybdate nanosheets onto nickel titanate nanofibers: Facile synthesis and efficient photocatalytic removal of tetracycline hydrochloride. *J. Colloid Interface Sci.* **2018**, *521*, 42–49. [CrossRef] [PubMed]
35. Li, S.; Hu, S.; Zhang, J.; Jiang, W.; Liu, J. Facile synthesis of Fe_2O_3 nanoparticles anchored on Bi_2MoO_6 microflowers with improved visible light photocatalytic activity. *J. Colloid Interface Sci.* **2017**, *497*, 93–101. [CrossRef] [PubMed]
36. Li, S.; Shen, X.; Liu, J.; Zhang, L. Synthesis of Ta_3N_5/Bi_2MoO_6 core-shell fiber-shaped heterojunctions as efficient and easily recyclable photocatalysts. *Environ. Sci. Nano* **2017**, *4*, 1155–1167. [CrossRef]
37. Korala, L.; Germain, J.R.; Chen, E.; Pala, I.R.; Li, D.; Brock, S.L. CdS aerogels as efficient photocatalysts for degradation of organic dyes under visible light irradiation. *Inorg. Chem. Front.* **2017**, *4*, 1451–1457. [CrossRef] [PubMed]
38. Zhang, Y.; Han, L.; Wang, C.; Wang, W.; Ling, T.; Yang, J.; Dong, C.; Lin, F.; Du, X.-W. Zinc-Blende CdS Nanocubes with Coordinated Facets for Photocatalytic Water Splitting. *ACS Catal.* **2017**, *7*, 1470–1477. [CrossRef]
39. Zhao, N.; Peng, J.; Liu, G.; Zhai, M. PVP-capped CdS nanopopcorns with type-II homojunctions for highly efficient visible-light-driven organic pollutant degradation and hydrogen evolution. *J. Mater. Chem. A* **2018**. [CrossRef]
40. Kandi, D.; Martha, S.; Thirumurugan, A.; Parida, K.M. Modification of BiOI microplates with CdS QDs for enhancing stability, optical property, electronic behavior toward rhodamine B decolorization, and photocatalytic hydrogen evolution. *J. Phys. Chem. C* **2017**, *121*, 4834–4849. [CrossRef]
41. Kaur, A.; Kansal, S.K. Bi_2WO_6 nanocuboids: An efficient visible light active photocatalyst for the degradation of levofloxacin drug in aqueous phase. *Chem. Eng. J.* **2016**, *302*, 194–203. [CrossRef]
42. Yang, X.; Xiang, Y.; Wang, X.; Li, S.; Chen, H.; Xing, D. Pyrene-based conjugated polymer/Bi_2MoO_6 Z-scheme hybrids: Facile construction and sustainable enhanced photocatalytic performance in ciprofloxacin and Cr(VI) removal under visible light irradiation. *Catalysts* **2018**, *8*, 185. [CrossRef]
43. Li, S.; Hu, S.; Jiang, W.; Liu, Y.; Liu, J.; Wang, Z. Synthesis of n-type TaON microspheres decorated by p-type Ag_2O with enhanced visible light photocatalytic activity. *Mol. Catal.* **2017**, *435*, 135–143. [CrossRef]
44. Chang, F.; Wu, F.; Zheng, J.; Cheng, W.; Yan, W.; Deng, B.; Hu, X. In-situ establishment of binary composites a-$Fe_2O_3/Bi_{12}O_{17}Cl_2$ with both photocatalytic and photo-Fenton features. *Chemosphere* **2018**, *210*, 257–266. [CrossRef] [PubMed]
45. Li, S.; Jiang, W.; Xu, K.; Hu, S.; Liu, Y.; Zhou, Y.; Liu, J. Synthesis of flower-like AgI/BiOCOOH pn heterojunctions with enhanced visible-light photocatalytic performance for the removal of toxic pollutants. *Front. Chem.* **2018**, *6*, 518.
46. Chen, Y.J.; Tian, G.H.; Shi, Y.H.; Xiao, Y.T.; Fu, H.G. Hierarchical MoS_2/Bi_2MoO_6 composites with synergistic effect for enhanced visible photocatalytic activity. *Appl. Catal. B* **2015**, *164*, 40–47. [CrossRef]

MDPI
St. Alban-Anlage 66
4052 Basel
Switzerland
Tel. +41 61 683 77 34
Fax +41 61 302 89 18
www.mdpi.com

Catalysts Editorial Office
E-mail: catalysts@mdpi.com
www.mdpi.com/journal/catalysts

www.ingramcontent.com/pod-product-compliance
Lightning Source LLC
LaVergne TN
LVHW070643170726
843515LV00004B/312

* 9 7 8 3 0 3 6 5 4 2 8 3 6 *